Principles and Practice of Bacteriology

Principles and Practice of Bacteriology

Paul Jaxon

MURPHY & MOORE
www.murphy-moorepublishing.com

Published by Murphy & Moore Publishing,
1 Rockefeller Plaza,
New York City, NY 10020, USA
www.murphy-moorepublishing.com

Principles and Practice of Bacteriology
Paul Jaxon

International Standard Book Number: 978-1-63987-451-4 (Hardback)

Cataloging-in-Publication Data

Principles and practice of bacteriology / Paul Jaxon.
p. cm.
Includes bibliographical references and index.
ISBN 978-1-63987-451-4
1. Bacteriology. 2. Microbiology. 3. Bacteria. I. Jaxon, Paul.
QR41.2 .P75 2022
579--dc23

TABLE OF CONTENTS

PREFACE

Bacteria are single celled microorganisms which lack nuclear membrane. They are metabolically active and divide using the techniques of binary fission. Bacteria are simple forms of life which have a highly adaptable nature. Pathogenic bacteria are those varieties of bacteria which are capable of causing infectious diseases in humans. The branch of biology which studies the various aspects of bacteria such as its morphology, genetics, ecology and biochemistry is known as bacteriology. It also attempts to identify, classify and characterize various bacterial species. The causative relationship between a disease and a microbe is established using Koch's postulates. Major advances in the field of bacteriology are the discoveries related to vaccines and antibiotics. This book presents the complex subject of bacteriology in the most comprehensible and easy to understand language. Most of the topics introduced herein cover the principles and practices of bacteriology. For all those who are interested in bacteriology, this book can prove to be an essential guide.

A detailed account of the significant topics covered in this book is provided below:

Chapter 1- Bacteriology deals with the study of bacteria and its related aspects. Morphology, ecology, structure, genetics, biochemistry etc. are a few of the facets that are studied under bacteriology. The identification, classification and characterization of bacterial species are also involved in bacteriology. This chapter provides an introduction to bacteriology in a simple manner.

Chapter 2- Bacteria are classified according to the nature of their cell walls, shapes or genetic makeup. Another way used to classify bacteria is Gram staining which further segregates them into Gram positive and Gram negative bacteria. This chapter explores the different types of bacteria to provide an extensive understanding of the subject.

Chapter 3- Bacterial genetics is the study of genetic materials of bacteria and their transfer during procreation. Through binary fission, bacteria divide into two daughter cells leading to bacterial reproduction. This chapter on bacterial genetics, bacterial growth and the various processes associated with the transfer of genetic material has been carefully written to provide an easy understanding of the topic.

Chapter 4- Bacteria contain a large number of proteins which are essential for the survival of the organism. They are arranged in network-like structures that perform distinct cellular functions. The various proteins found in bacteria like the AB toxin, cholera toxin, lac repressor, etc. are methodically elaborated in this chapter to help the reader gain a better perspective of the subject.

Chapter 5- While most bacteria are useful, some cause various illnesses and diseases in other organisms. Bacterial diseases occur when these harmful bacteria get into a body and crowd out the healthy ones. Some bacterial diseases are bacterial pneumonia, actinomycosis, leprosy, tuberculosis, etc. This chapter discusses these bacterial diseases and their causes in detail.

It gives me an immense pleasure to thank our entire team for their efforts. Finally in the end, I would like to thank my family and colleagues who have been a great source of inspiration and support.

Paul Jaxon

Chapter 1
Understanding Bacteriology

Bacteriology deals with the study of bacteria and its related aspects. Morphology, ecology, structure, genetics, biochemistry etc. are a few of the facets that are studied under bacteriology. The identification, classification and characterization of bacterial species are also involved in bacteriology. This chapter provides an introduction to bacteriology in a simple manner.

Microbiology

Microbiology is the study of biological organisms that are too small to be seen with the naked eye (without using such tools as the magnifying glass or microscope etc). Microbiology is therefore dedicated to studying the lives and characteristics of a wide variety of organisms ranging from bacteria and archaea to parasitic worms in their environments. The discipline is used to learn about all aspects of the organisms in order to not only determine how they live in their environment, but also how they impact their respective surroundings and thus other organisms around them (human beings, animals, etc).

Microbiology has proved to be one of the most important disciplines in biology, making it possible to identify how some of these organisms cause diseases, discover cures for such diseases and even use some microbes for industrial purposes etc.

Microbiology can be divided into two branches: pure and applied. The former is the most fundamental branch, in which organisms themselves are examined in-depth. In applied microbiology, the organisms themselves are not studied but are applied to a certain process. Additionally, microbiology can be separated by taxonomy, which classifies different microbes (bacteriology, protozoology, etc).

Pure Microbiology

- Microbial Genetics: Studying the genetics within microorganisms, including bacteria, viruses and fungi, is the principal focus of microbial genetics. The examination of their growth rates and generation cycles helps scientists better understand a microorganism's evolution.
- Immunology: Concentrating on the study of an organism's immune system is the specialty of immunology. Scientists conduct research on the immune

system to treat disorders within it, including autoimmune diseases, transplant rejection and hypersensitivities.

Applied Microbiology

- Medical Microbiology: Focusing on the application of medicine, medical microbiologists diagnosis, treat and prevent infectious diseases. Additionally, these scientists study microbes to enhance overall health in humans.
- Biotechnology: Investigating methods of using living organisms to invent or produce products is the specialty of a biotechnologist. By using genetic engineering, along with cell and tissue technologies, physicians can customize treatments for their patients.

Taxonomy

- Bacteriology: Researching methods related to producing medicine from bacteria is the study of bacteriology.

Bacteriology

Bacteriology is a branch or discipline of science that studies different characteristics of bacteria and their association with other organisms or disciplines. Over the years, the discipline of bacteriology has evolved from the microbiological tests performed only by physicians, the application of the germ theory of disease and economic concerns relating to the spoilage of foods and wine to the use of bacteria to simpler forms of laboratory tests and the production of new nutritious food and antibiotics.

The study of bacteria began with the discovery of the microscope and was mostly limited to experts and scientists of that time. The initial advances in pathogenic bacteriology began with the identification and characterization of bacteria associated with specific diseases. Nowadays, however, bacteriology is used in various other fields like food, pharmaceuticals, agriculture, and the environment.

Significant advances made in bacteriology over the last century have resulted in the development of many useful vaccines like pneumococcal polysaccharide vaccine, diphtheria toxoid as well as antibiotics.

At present, most bacterial diseases in humans and their infectious agents have been identified, although significant variants continue to evolve and sometimes emerge. Studies made in bacteriology have also allowed the use of gene recombination technology in bacteria to produce improved bacterial strains that can then be used in various areas. The importance of bacteriology continues to grow throughout the years as many diseases and disorders associated with almost all living beings are caused by bacteria.

Studies of non-pathogenic bacteria that exist in a symbiotic relationship with plants and animals are even beneficial to said organisms. The identification and classification of bacteria into different groups have allowed for a convenient way of studying these bacteria and the formation of a separate branch, systemic bacteriology that deals with taxonomy. Besides, bacteriological studies, subsequently, have also been divided into different areas like soil bacteriology, public-health bacteriology, industrial bacteriology, etc.

Bacteriological studies have been the focal point for disease diagnosis and treatment, and their importance continues to grow with the emergence of new bacteria and associated diseases.

Medical Bacteriology

Medical microbiology, also known as clinical microbiology, is a sub discipline of microbiology dealing with the study of microorganisms (parasites, fungi, bacteria, viruses, and prions) capable of infecting and causing diseases in humans. As with other specialties in medicine, the evolution of medical microbiology is essentially a response to clinical needs. At its core is the study of the pathogenesis and epidemiology of these microorganisms, and its practical applications lie in the form of diagnostic and therapeutic guidance – intrinsically a supportive role in clinical practice.

Clinical microbiology and bacteriology in particular has always been a slowly evolving science. Technologies developed in the 19th century still play a central role in routine diagnostics and subsequent therapeutic responses. Classical growth-based microbiology similar to that employed by Pasteur and Koch is not only used today for microbial detection and identification, it also provides the technological basis for much of the antimicrobial susceptibility testing (AST) that is performed. In addition, even older technologies, including a form of light microscopy that would be recognizable by Van Leeuwenhoek himself, continue to be commonplace. Figure below summarizes today's workflow in this conservative environment. In the bacteriology laboratory, incoming samples are usually subjected to Gram-staining and microscopy after which the same materials are prepared for cultivation on solid or in liquid growth media. After incubation for a sufficient duration at an appropriate temperature in a suitable atmosphere, microbial growth is evaluated. Samples without microbial growth are discarded whereas positive samples are evaluated further including more detailed strain characterization (often initiated by a second and time-consuming round of purifying cultivation) using simple and inexpensive testing (motility assessment, catalase testing etc.). This leads to (partial) species assessment, which may be completed using manual or automated biochemical testing and extended enzymology. Next or sometimes even simultaneously, the bacterial isolates can be further characterized with respect to their antimicrobial susceptibility profiles and their epidemiological characteristics. For these studies, many classical and some automated basic microbiology methods have been made available. This overall picture essentially describes the core activities of a clinical microbiology laboratory as it has functioned for the past century.

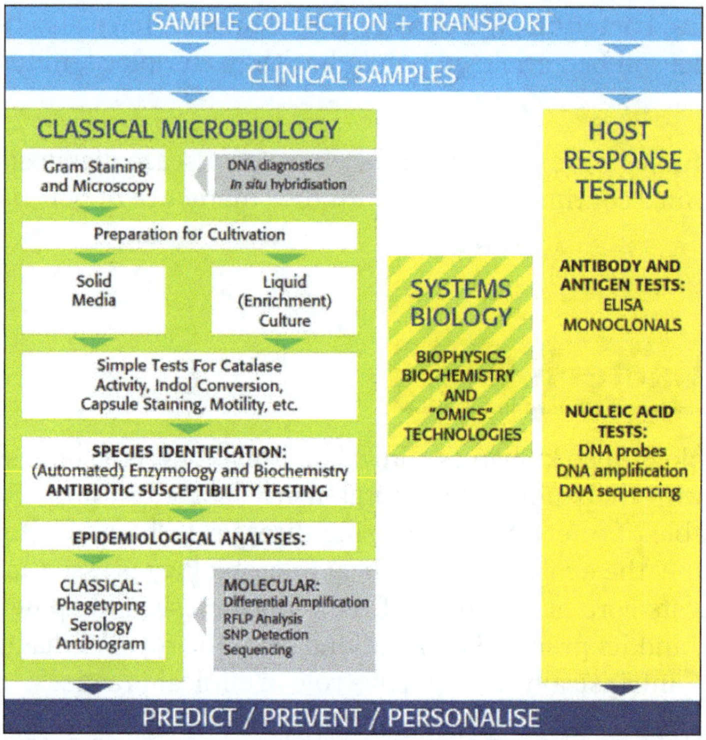

More recently, several waves of mostly technologically oriented innovation have been observed. With the advent of immunological methods, measurements of host reactivity to infection were facilitated. Definition of antibody specificity and levels using ELISA, co-determined by the exquisite specificity of monoclonal antibodies, became an accepted diagnostic standard for the detection of many microbial infections. Similar serological tests have also been used for epidemiological characterization of microorganisms. After the establishment of immunodiagnostics the development of molecular-based testing was not far behind. Although primarily developed in research oriented laboratories, molecular technology has acquired an important place in clinical microbiology. The direct detection of nucleic acids specific for certain microbial species has revolutionized, for instance, the detection of the sexually transmitted bacterial species Chlamydia trachomatis and Neisseria gonorrhoeae. The use of DNA probes and nucleic acid amplification became well accepted although the analytical specificity and sensitivity of some of these tests may still require optimization. Today biophysical technology is entering the clinical microbiology arena (e.g. matrix-assisted laser desorption ionization time of flight mass spectrometry (MALDI-TOF MS)) and this will allow for further automation of laboratory procedures.

Hence, over the past two decades clinical microbiology laboratories have been genuinely transformed. Direct host testing became more integrated and molecular and biophysical diagnostic technologies were successfully introduced. Over the coming years, the sequential or combined introduction of aspects of the other "omics" technologies (genomics, transcriptomics, proteomics, glycomics, lipidomics etc.) will further enhance the implementation of real-time rational therapies and the improvement of microbiological surveillance. This approach will, in turn, lead to prediction, prevention and personalization of the infectious risk assessment and, hopefully, more efficacious treatment of infectious diseases.

Clinical Specimens

Clinical specimens and their management are key quality determinants in clinical microbiology. From the moment a specimen is collected, timing becomes critical for microbial viability i.e., the longer specimens are held under bacterial growth-limiting conditions, the smaller the chance of recovering microorganisms by growth-based methods. This, in turn, can lead to drastic performance changes in test sensitivity and specificity as well. Strictly anaerobic organisms, for instance, may ultimately be overgrown by even small numbers of the conditional anaerobic or fully aerobic species. This means that speed of transportation is an important quality parameter in clinical microbiology. Unfortunately the clinical laboratory often cannot stringently control this aspect of quality assurance other than outright rejection of poorly handled clinical materials. Policies to improve awareness are very important in this respect. From a research and development point of view there is an obvious need for improvement of transport media when organism viability is essential. Since some bacterial species are considered to be uncultivable on the presently used artificial growth media, there is also an obvious need for designing new media or alternative culture formats. When diagnostics involve the detection of DNA or other cellular components, a good lysis buffer containing compound-specific stabilizers (e.g. DNase or protease inhibitors) is equally important. The diverse nature of clinical specimens including the biological diversity normally encountered even in different samples of the same clinical material renders adequate detection and quantification a complex and challenging task. Equipment that would facilitate the parallel purification of host cells, bacteria, nucleic acids and/or viruses and proteins or other sub-cellular elements from diverse clinical materials would certainly be both clinically and commercially successful. Obviously, the developments of methods that utilize direct enumeration of potentially pathogenic microorganisms using stabilized nucleic acids in samples are less susceptible to transport stringency.

Culture is not Dead

Culture has been the mainstay of clinical microbiology for the past century and will likely remain so for decades to come despite accelerating technological development and the introduction of novel diagnostic procedures. However, there remains an absolute need for significant quantities of living organisms - not as much for detection and identification of microbial species, since strong alternative technologies are being introduced including MALDI-TOF MS, but more specifically for AST. The ability to document the killing or stationary effect (or lack thereof) of antibiotics on intact, living cells has not yet found a suitable growth-free alternative although advances toward this goal are currently under development. Bacterial killing by antibiotics is usually monitored by the visible changes in bacterial density in liquid or solid growth media in the absence or presence of strategic concentrations of those antibiotics with clinical relevance. The read-out technology may change in the coming years but straightforward live-dead monitoring is still at the core of AST until bacterial death can be reliably and reproducibly measured in single cells or small numbers of cells in a culture-free environment. To date, most applications require between 104 and 105 microbial cells per assay.

However, classical culture is also being challenged and optimized continuously. The standard Petri dish format is faced with competing films, fibers and nano-porous carrier materials in which the actual "growth containers" may have internal volumes in the nano-liter range. This enables development of contained growth and/or micro-monitoring on alumina oxide chips for example. Solid agar media can be loaded with chromogenic compounds that are specifically metabolized by a certain bacterial species only. This facilitates a direct on-the-plate, visual species identification that hastens detailed reporting of culture results. When antibiotics are included in the chromogenic or standard agar-based media, it might provide presumptive species identification but also limited resistance screening. It must be emphasized that AST based upon growth inhibition on solid media is still immensely popular. Disk diffusion testing and/or EtestR are still used as a primary means of AST in many laboratories. The automation of solid media inoculation is becoming extremely important for large sample volume laboratories. A number of systems are currently marketed for that purpose including mechanisms for automated incubation and overall management of actively growing cultures. Such machines usually harbor cameras for real time archiving of (growing) cultures, demonstrating that detection and morphological characterization of (micro-) colonies may potentially speed up time to results for culture-based technology, certainly if growth could be monitored from a distance (e.g. via an internet connection with the incubator's data acquisition station). It has to be noted that such requirements for small versus big laboratories or laboratories in developed versus developing countries may be hugely different.

Liquid culture is continuously optimized as well - not only by changing the medium composition but also by progressive automation and the use of increasingly sophisticated growth sensors. Special compounds, such as resins that specifically capture and inactivate antibiotics present in patient samples, are added to reduce the time to positivity. AST in liquid phase is easily automated and a number of systems are available for the clinical microbiology laboratory including the Becton Dickinson Phoenix, the Siemens Micoscan WalkAway or the bioMérieux VITEK2™ systems, which are the three best-known examples.

It has to be noted that there are no studies that unambiguously demonstrate the added value of "rapidity adaptations" of classical microbiology methods. This has to be realized by those involved in microbiological R&D and the focus should clearly remain on the acceleration of these improvements. Obviously, alternative approaches rather than classical methodology may then be obligatory.

Current Trends Towards Improvement

Clinical microbiology is evolving in a speed that is much higher than it used to be in the previous century. As a consequence, the availability of experienced laboratory personnel is becoming a critical issue and although properly trained technicians are rare, there is a clear need to adapt their training to meet the diagnostic expectations. Current developments in clinical microbiology, in one way or another, target laboratory automation, high level information generation and reducing the overall time to results of detection, identification, and AST of bacteria, yeasts and molds in general. Secondary trends are related to connectivity (between automated systems, between systems and the laboratory or hospital information systems (LIS, HIS) and between the clinical microbiologist and the physician), cost effectiveness, quality services and clinical information content (how to deliver the most useful microbiological information as quickly as possible to clinicians). As technological innovations, imaging, mass spectrometry (MS) and sequencing are at the core of diagnostic R&D and so too are possible improvements in the speed, accuracy, accessibility and clinical correlation of AST results.

Molecular Testing

Nucleic acid testing was introduced in the 1980's. Plasmid purification and (in situ) nucleic acid probe-mediated detection, identification and characterization were the first technologies developed. Neither of these was broadly accepted by the clinical community. The same happened to restriction fragment length polymorphism (RFLP) analysis and a few other even more obscure molecular test formats. However, epidemiological typing using pulsed field gel electrophoresis (PFGE) of DNA macro-restriction fragments became an undisputed gold standard in many laboratories and public health laboratories in particular. And of course sequencing and PCR became much more appreciated over the past decade. PCR by itself has revolutionized the science of clinical virology generating numerous novel assays with an obvious impact on the care of immunosuppressed and transplant patients. Cell culture-based assays for C. trachomatis were fully replaced by amplification and for all clinically relevant microorganisms multiple sensitive and specific PCR tests were developed, some of which are now fully automated. Real point-of-care tests for difficult-to-detect organisms including Mycobacterium tuberculosis are currently available, generating a result within the hour without any complicated sample processing steps. Even the difficult-to-diagnose syndromes such as bacterial sepsis and pneumonia can be addressed using some form or another of molecular testing. In addition, specific tests for a variety of virulence or anti-microbial resistance genes have been developed in large numbers. Several of these tests have been cleared by the American Food and Drug Administration (FDA) and are hence ready to fulfill their central role in laboratory innovation. Clearly, PCR is here to stay, although broad acceptance of many of the PCR tests has not yet been fully realized. This also includes tests that monitor host polymorphisms that can predict risk of colonization and/or infection. Such assays will undoubtedly advance and expand in number and quality over the coming decade. Technological improvements are continuous: the use of droplet PCR, for instance, allows for very accurate quantification of the number of target molecules in the clinical starting material. The availability of so-called aptamers, single stranded short RNA or DNA molecules that have exquisite affinity for only a single ligand may again significantly impact nucleic acid diagnostics.

Mass Spectrometry

MS was recently introduced as a new diagnostic tool in clinical microbiology with a first successful application for the detection and sequence-based identification of PCR products. Although the biophysical technology is intrinsically complicated, it has to be said that from a practical, laboratory-utilization perspective MALDI-TOF MS is quite simple from a performance standpoint. A bacterial colony grown on a solid agar medium needs to be carefully sampled with a calibrated plastic loop or a toothpick. The biomaterial is deposited on a sample holder and a matrix solution needs to be added. Simple and direct pre-spotting lysis protocols for yeasts, molds and mycobacteria are also available. The sample slide is positioned in the MS machine and bacterial species identification is accomplished by comparing the generated spectrum of proteins (and peptides) to defined mass spectra contained within a reference spectral database representing multiple strains of known bacterial species. The current databases contain hundreds to thousands of such reference spectra and species identification (ID) is reliable and fast. This is precisely why this technology was so quickly accepted by clinical microbiologists. After the introduction of serological and molecular methods, biophysics has now definitely and irreversibly entered the clinical microbiology laboratory.

MALDI-TOF MS is extremely well suited for species identification since the major mass profile that is generated comes primarily from ribosomal proteins, which aligns nicely with current taxonomic classification. However, other applications are being realized including epidemiological typing which is relatively easy to perform relative to other standard methods such as PFGE and multi-locus sequence typing (MLST). For the purpose of epidemiological typing of bacterial isolates, mass peaks that are specific for a given microbial strain are easily identified and can be used to develop binary typing systems. In addition, certain resistance mechanisms can be identified as well. In case of AST or antimicrobial resistance testing (ART), degradation products of beta-lactam antibiotics that were hydrolyzed by beta-lactamases have been identified in a number of independent studies. By using different (mixtures of) beta-lactam antibiotics as substrates, adequate identification of beta-lactamases could be achieved. Cellular changes or modifications under the influence of antibiotics can be detected as well as was recently demonstrated for the interaction between Candida albicans cells versus fluconazole and Candida and Aspergillus species versus caspofungin. In addition, MS can also be used to identify PCR products derived from resistance genes. The iPLEX MassArray assays developed by Sequenom (San Diego, CA, USA) provide a good example of this flexible technology. The use of reporter molecules for which the mass changes when in contact with microbial virulence factors has been described to help assess the putative invasive potential of bacterial species. Signal peptides that are specifically cut by known proteases have shown diagnostic value for the identification of anthrax and periodontitis.

Of course there are numerous confounding issues associated with diagnostic MS. First, culture is required since the sensitivity of the systems is in the order of 104-105 cells per assay. These are numbers rarely encountered in clinical specimens, with the possible exception of patients with significant urinary tract infection. Second, equipment is expensive, bulky, sometimes noisy and in need of costly maintenance. Third, the distinction between certain genera, species or pathovars (Escherichia coli and Shigella spp for example) is still hard to make. This renders regular updates of the database a strict requirement.

(Next Generation) Nucleic Acid Sequencing

DNA sequencing had already developed into a standard research tool before it entered into the clinical arena. The technology as developed by Sanger is still widely used to sequence relatively short stretches of (usually PCR-amplified) DNA. This has been particularly useful for 16S rDNA sequence identification of bacteria at the species level, but also for epidemiological purposes. Sanger sequencing is frequently used to detect point mutations associated with resistance to antiviral compounds or antibiotics. However, the importance of sequencing became even more apparent with the advent of newer generation sequencing procedures in infectious disease diagnosis. These methods for the first time allowed for full sequencing of entire bacterial genomes. When whole genome sequencing is applied in the context of microbial epidemiology or antimicrobial resistance, it can help in generating refined dissemination maps as well as "resistomes" and "toxomes".

To date, the technology has advanced to a state where even the genomes of complex mixtures of bacteria can be determined. This so-called microbial metagenomics can be used to generate complete catalogues of genomic components of bacterial species present in a variety of clinical materials including fecal specimens. In turn, this enables the R&D laboratory to detect differences in the composition of the bacterial flora in healthy versus diseased individuals. This will surely identify species that may be causal in syndromes as different as asthma, Crohn's disease and cardiovascular afflictions, although to date it is not clear when this technology will become routine in the clinical microbiology laboratory. Metagenomics is essential in trying to understand the overall effect of antibiotic treatment on the human microbiome as a whole. At a more advanced state, next generation sequencing (NGS) will also facilitate genetic testing of host susceptibility towards infection or colonization.

The commercial availability of DNA sequencing instrumentation is already large. Companies such as Illumina, Roche and Ion Torrent offer affordable laboratory equipment with a limited footprint and simplified physics and chemistry. The fact that some of these technologies in the end will also enable sequencing of additional biopolymers (proteins, oligo-and polysaccharides and fatty acids) and probably even detection and identification of small molecules of a diverse chemical nature will result in even more accelerated market penetration of unprecedented diagnostic assays.

Alternative Technologies

In addition to the more classical diagnostic microbiology approaches there are a number of novel technologies that may be of significance in the not-too-far-away future. Some of these are highlighted below and their potential diagnostic relevance is discussed.

Electronic Nose Devices

The problem with this approach has always been the limited reproducibility of the diagnostic assays and the difficulties encountered in concentration and/or efficient capture of the volatile compounds to be detected and/or identified. In addition, the equipment required was and generally still is complicated to work with, bulky and expensive. A cheap alternative (depending on eating habits) may depend on animals that can differentiate specific scents although that is still considered controversial: having bulky rats with 50 cm tails perform clinical investigations will change the perception of "clinical rounds" in general. Hence, many diagnostic tests have been described but essentially none of them have attained routine status. However, the promise of direct breath testing of infected individuals could open innovative diagnostic avenues in the future.

The most widely used VOC detection procedures encompass micro-weighing techniques using vibrational methods, gas chromatography and changes in electrical conductivity of metal chips. The latter technology has recently been improved significantly with the design of systems that allowed for the kinetic measurement of VOC production by growing bacteria in the head space of an enclosed culture container. Using such approaches it has been demonstrated that the distinction of several bacterial species could be reliably made on the basis of their longitudinal VOC production profile. Next to this dynamic and culture medium-dependent method, it also appears that direct assessment of disease specific tuberculosis markers is feasible. A field study in Bangladesh, among patients and closely matched controls revealed that patients could be identified with reasonable sensitivity and specificity (manuscript submitted). It was anticipated that a diagnostic tool could be produced for less than 100 euro, whereas the tool could be regenerated and used for the diagnosis of multiple patients.

This would suggest that the detection of VOCs could be developed into point of care assays that would possibly enable improved treatment procedures for those patients in geographic regions where this kind of care is most needed. The combined uses of ion flow tube MS and thermal desorption gas chromatography MS helped identify 2-phenylethanol as a unique volatile marker for mycobacterial growth. This work could potentially be translated into specific test for this or other organism-specific volatiles. In addition, these formats do show promise for antibiotic susceptibility testing in bacteriology.

Vibrational-and Absorption-based Spectroscopy

There are two examples of vibrational/absorption-based technologies that have been addressed with equal frequency. Infra-red spectroscopy depends on irradiation of biological specimens with infra-red light followed by measurement of characteristic absorption and transmission profiles. The disadvantage of this technology is that water strongly contributes to the absorption spectra, which complicates the analysis. This may be why Raman spectroscopy has been favored slightly. Raman spectroscopy measures light scattering that takes place upon illumination of biological specimens with visible (laser) light. The disadvantage in this case is that autofluorescent bacterial components can strongly interfere with spectrum generation (e.g. carotene in Staphylococcus aureus). A relatively pure preparation of bacteria containing a significant number of cells is required for either technology, although single cell applications have been described. We will only mention some examples of Raman applications, but we emphasize that infrared (IR) spectroscopy may be equally applicable if the water signal is efficiently suppressed.

The most "visible" contribution of Raman spectroscopy in clinical microbiology has been in the field of microbial epidemiology. Although it has repeatedly been claimed that Raman spectroscopy can be used to distinguish virtually all clinically relevant bacterial species, definitive data are thus far missing. However, several studies concluded convincingly that typing of different strains belonging to a single species can be performed adequately. For various species of Staphylococcus, Escherichia coli, Klebsiella pneumoniae and Pseudomonas aeruginosa, it has been demonstrated that Raman-mediated typing generates results that are concordant with those of current gold standards. This has shown usefulness for delineating nosocomial outbreaks and even larger scale global dissemination of certain multi-drug resistant strains further suggesting that Raman (and IR) spectroscopy could be developed into useful diagnostic tools with additional research.

A variety of (biophysical) methods have been introduced into the clinical microbiology arena over the past decade. These methods clearly meet certain diagnostic requirements and development is surely continuing. In addition, the use of microfluidic tools will lead to miniaturization of equipment and more facile single cell handling. Such systems, when simplified to paper and lateral flow formats with targeted capillary capabilities, may be well suited for (low budget) field studies even in developing countries. Some of these methods have been successfully introduced in the clinical microbiology laboratory which remains a matter of time for numerous other applications. For other methods, e.g., impedance sensing, there is only a proof of principle study at present with no information regarding feasibility. Some studies, however, have demonstrated useful results for both gram positive and negative bacterial identification. Evolution of detection and identification assays in clinical microbiology progresses swiftly.

Immediate Urgencies

The development and spread of multi-drug resistant microorganisms including pan-resistant strains generates a universal threat to both humans and animals. Prevention of the development of pan-resistance seems impossible given the environmental antibiotic-mediated selective pressure. Because of this, the diagnostic microbiology field should be strategically aligned towards surveillance and early detection of such resistance. However, AST is lagging behind in continuous improvement as compared to microbial detection and identification. Where most microbiologists will agree that nucleic acidbased diagnostics and MS will dominate the detection/identification fields over the coming years, the prospects for AST improvements are relatively ill-defined. To date, AST is primarily manual, using classical tests such as the use of antibiotic-containing screening media, disk diffusion methodology or EtestR. While automation is available (VITEK2™, Phoenix, Microscan WalkAway etc.), the technology remains growth-dependent. Of course, with the availability of PCR, several systems for molecular AST have been proposed. Still, the "real time PCR antibiogram" is a research rather than a diagnostic tool. Again, this still remains stuck in the research phase. For future generation methodologies, the experts seem to be agreeing that NGS and also "next generation" MS may help to develop this analytical field, but the time scales for implementation are not clear. It is for these reasons that "intermediate technologies", i.e., ones that can be implemented sooner and provide real progress with regard to the classical culture-based methods are eagerly awaited. Below and in random order, are a limited number of technologies that may in the end help solve the current status quo.

Imaging for AST

AST is a core component of clinical microbiology since it guides proper antimicrobial chemotherapy. Obviously, many simple tools for facilitating its interpretation have been developed. One of them is simple photography of plates where disk diffusion tests were applied. Camera systems have been commercialized and these show useful efficacy. This was extended for use with micro-colonies as well. Serial photography generated mini-movies of (non-)growing micro-colonies in the selective presence or absence of various antibiotics. The BACcel™ system as manufactured by Accelr is a good example of a machine that can do a one day AST for bacteria grown in blood culture bottles. It shows that simple test formats can still be adapted into systems that serve the clinical microbiologist's needs in a better way.

Fluorescence Activated Cell Sorting

Fluorescence activated cell sorting or FACS enables distinction of cells with different sizes and distinct levels of fluorescence. The fluorescence can be intrinsic to the cells or it can be specifically attached to or introduced in subsets of cells. For the first application, fluorescent antibodies are frequently used. There are also fluorescent stains that penetrate only dead cells or cells with a permeable membrane. Hence, cells that are affected or killed by antibiotics can potentially be distinguished from viable ones. During susceptibility testing, the ratio between dead and/or non-dividing cells changes over time and the use of fluorescent labeling and FACS can then help to determine the bactericidal or bacteriostatic effect of the antibiotic studied. It has been shown that methicillin susceptible S. aureus (MSSA) and methicillin resistant S. aureus (MRSA) can be distinguished by FACS after two-hour incubation with oxacillin. When using micro-fluidic methods in combination with FACS, one can also distinguish cellular shapes and sizes. Although this has only been explored for red blood cells. Obviously, this technology is still far away from direct clinical application.

Microbial Cell Weighing by Vibrating Cantilevers

Cantilevers containing small canals which facilitate bacterial passage can be made to vibrate continuously. When bacteria pass such a system their sheer weight (in the femtogram range) will cause a change in the frequency of cantilever movement. The nature of the change is correlated with the weight of the passing cells and "light" cells will cause a change in vibration that differs from that induced by "heavy" cells. When cells are treated with certain antiseptics or antibiotics or when they are subjected to unfavorable environmental conditions such as osmotic shock, their buoyant mass density may change and these changes can be measured. This has been demonstrated for ampicillin resistant and susceptible variants of Citrobacter rodentium. It was also shown that resuscitation after osmotic shock for both variants and in the presence or absence of ampicillin could be easily differentiated in a short time-span. The method could potentially allow determination of the growth rate of bacterial cells having a doubling time of ten hours in a mere ten minutes. Cantilevers can be multiplexed using nanotechnology such that multiple antibiotics in various concentrations could be tested for a single growing culture simultaneously.

Micro-calorimetry

Bacteria produce energy and the level of production is affected by the presence of antibiotics to which the microbes may be susceptible. In case of resistance, steady-state energy production will be different in the presence of antibiotics. Recent studies in tuberculosis have shown the usefulness of isothermal micro-calorimetry for the rapid detection of mycobacteria and their susceptibility to isoniazid, ethambutol, and moxifloxacin. The maximum bacterial growth rate and the lag phase were quantified by integrated heat flow-versus-time analysis. The micro-calorimetric technology used consisted of broadly affordable and sensitive micro-calorimeters. Also using isothermal micro-calorimetry, bacterial species identification from urine specimens was suggested to be possible even at low bacterial counts within a little over three hours on the basis of dynamic heat flow patterns.

Chip calorimetry is a monitoring tool for determining the physiological state of biofilms. Its potential use for the study of the effects of antibiotics was tested using an established model. The real-time monitoring potential of chip calorimetry was successfully demonstrated: a dosage of antibiotics initially increased the heat production rate probably due to activity of energy-dependent resistance mechanisms. The subsequent reduction in heat production was probably due to the loss of activity and the death of the biofilm bacteria. This new analytical tool provided fast, quantitative, and mechanistic insights into the effects of antibiotics on biofilm activity. Detailed studies on the usefulness of this technology on AST in general are ongoing in various institutions worldwide.

Listening at the Microscale

Using optical tweezers capturing little gold nanoparticles and a microscopic sound source, it was recently shown that differences in the vibrational and, hence, energy level of the gold particles could be efficiently measured in liquid media. The gold particle served as a specific "nano-ear" in this model system. Replacing the gold particle by a bacterial cell, either resistant or susceptible to a given antibiotic, in liquid media with or without antibiotics would facilitate characterization of the energy state of the single cells which is supposed to vary on the basis of phenotype and the presence or absence of antibiotics. The latter has not yet been convincingly demonstrated but further developments are eagerly awaited since this would provide another method for measuring the reactivity of single bacterial cells towards different antibiotics.

Rotating Magnets

When magnetic beads are brought into specific magnetic fields they adapt a specific rotational spin. The frequency of rotation can be influenced by the binding of other molecules, viruses or bacteria. So if the beads are equipped with a ligand that specifically captures bacterial cells, the rotation of the beads changes at the moment of capture. This change can be measured. If one would pair all beads in a broth culture with one or two cells, which can be done by incubating derivatized beads with a diluted bacterial suspension and then wash, they will assume a constant rotational frequency. If the bacteria start to divide, the rotation frequency starts to change. If this division can be inhibited or blocked by antibiotics (that is, if the cells are susceptible to the antibiotic applied) then the change will not take place. If the bacteria are resistant to the (amount of) antibiotic applied, then again the change in rotational frequency will take place. In this way bacterial resistance can be determined and precisely quantified. Interestingly, specific formats of magnetic nanosensors can also be used to diagnose the presence of intracellular pathogens in host cell populations. Novel developments in this field comprise the use of self-assembling magnetic particles that further facilitate efficient MIC measurements.

RNA Sequencing

In the recent past it has been demonstrated that pairs of bacterial strains that differ in AST profiles can be distinguished on the basis of the transcriptome generated in the presence or absence of antibiotics. Antibiotic exposure induces a stereotypical change in transcriptional response within a few minutes and genes involved in for instance SOS response were significantly up- or down-regulated. Even more recent data show that similar transcriptional changes can also be monitored for bacteria present in clinical materials. Again, antibiotic pulsing revealed specific transcriptional signatures for resistant versus susceptible bacteria and mecA or vanA transcription could be detected for MRSA and vancomycin resistant enterococci (VRE), respectively. Still, these investigations are in early developmental stages and further validation needs to be performed. The downside of this approach is that the presence of heteroresistant subpopulations might not be detected amongst the SOS signals generated by the susceptible population.

Testing in Microdroplets

Little milli-, micro- or nano-droplets can be used as small, individual reactors. When the droplets can be individually manipulated and when they contain bacteria in sufficient numbers of cells, these microenvironments can serve as mini-fermenters where metabolic activity and viability of cells can be monitored. Development of such a system became feasible once the emulsification process was controlled and the long term kinetic stability of the droplets could be guaranteed. The system consisted of 100 nL droplets containing 103 bacteria per droplet and in different droplets, different concentrations of antibiotics could be established. By following the droplets over time by epi-fluorescence measurement, survival or death could be monitored for the bacteria populating the different droplets. Even more recently droplets were prepared that contain single bacteria. This technology can be miniaturized, easily multiplexed with respect to the number of antibiotics tested per bacterial strain and duration of the test can be as short as a single or a few bacterial replication cycles. Obviously, assessing technical reproducibility and the development of adequate reference MIC databases will take many years.

Application of Bacteriophage Susceptibility

Bacteriophages need living bacteria for their replication such that bacteria that are susceptible to certain antibiotics cannot support the replication of bacteriophage in the presence of these antibiotics. In 1997, the first bacteriophage-based AST system was presented for Mycobacterium tuberculosis. Two years later a clinical validation study of six different antibiotics was published and it was demonstrated that a 90% drop in phage counts could be used as an indicator for the MIC. The system was optimized by including recombinant phages containing luciferase genes. Instead of doing phage counts, simple luminescence assays were developed. It was shown that the luciferase assay could be done at low costs in two days as opposed to the average of 10 days for classical AST for M. tuberculosis. These data were independently replicated and technological innovation continued. Some of the tests were even applied successfully in developing countries. The system was even further refined by using "fluoromycobacteriophages", phages with intrinsic fluorescence or by the application of real time PCR detection of (non-)replicating bacteriophages. Interestingly, the system was never widely accepted in clinical microbiology, nor was it expanded to other bacterial applications. Two of the major problems were laboratory contamination with bacteriophage and possible phage resistance of certain bacterial strains. So irrespective of the sound biological principle of this approach, this system is unlikely to change our AST practice.

Classical microbiological methods for AST can still be successfully applied but then become the time limiting step in the clinical microbiology laboratory. Innovation in the field of AST in particular supported by microfluidic approaches is urgently needed to accelerate clinical testing and improve medical decision making with regard to antimicrobial therapy. Obviously, genome sequencing will be considered for novel generations of AST formats and the first examples linking genome sequencing with the identification of appropriate antimicrobial therapy has already been published. As another example, programmed cell death in bacteria subjected to antimicrobials, highly sophisticated methods including nuclear magnetic resonance (NMR) or assessment of dielectric permittivity may help identify novel metabolic or biophysical markers of susceptibility or resistance. However, as for all of the other innovative methods, setting up optimal reference databases will take significant R&D efforts for the years to come.

Bacterial Taxonomy

The science of classification of bacteria is called bacterial taxonomy. Bacterial taxonomy (G: taxis = arrangement or order, nomos = law or nemein = to distribute or govern), in a broader sense, consists of three separate but interrelated disciplines: classification, nomenclature, and identification.

Classification refers to the arrangements of bacteria into groups or taxa (sing, taxon) on the basis of their mutual similarity or evolutionary relatedness. Nomenclature is the discipline concerned with the assignment of names to taxonomic groups as per published rules. Identification represents the practical side of taxonomy, which is the process of determining that a particular isolate belongs to a recognized taxon. It is to mention here that the term Bacterial systematics often is used for bacterial taxonomy.

But, systematics bears broader sense than taxonomy and is defined by many as the scientific study of organisms with the ultimate object of characterizing and arranging them in an orderly fashion. Systematics therefore encompasses disciplines such as morphology, ecology, epidemiology, biochemistry, molecular biology, and physiology of bacteria.

Importance of Bacterial Taxonomy

Bacterial taxonomy, however, is important due to following reasons:

- Bacterial taxonomy senses to be a library catalogue that helps easily access large number of books. Taxonomy therefore helps classifying and arranging bewildering diversity of bacteria into groups or taxa on the basis of their mutual similarity or evolutionary relatedness.
- The science of bacteriology is not possible without taxonomy because the latter places bacteria in meaningful, useful groups with precise names so that bacteriologists can work with them and communicate efficiently.
- Bacterial taxonomy helps bacteriologists to make predictions and frame hypotheses for further research based on knowledge of identical bacteria. For convenience, the bacteriologist can predict that a bacterium in question would be possessing similar characteristics to its relative bacterium whose characteristics are already known.
- Contribution of bacterial taxonomy in accurately identifying bacteria is of practical significance. For convenience, bacterial taxonomy contributes particularly in the area of clinical microbiology. Treatment of bacterial disease often become exceptionally difficult if the pathogen is not properly identified.

Ranks or Levels of Bacterial Taxonomy

In bacterial taxonomy, a bacterium is placed within a small but homogenous group in a rank or level. Groups of this rank or level unite creating a group of higher rank or level. In bacterial taxonomy, the most commonly used ranks or levels in their ascending order are: species, genera, families, orders, classes, phyla, and domain.

Species is the basic taxonomic group in bacterial taxonomy. Groups of species are then collected into genera (sing, genus). Groups of genera are collected into families (sing, family), families into orders, orders into classes, classes into phyla (sing, phylum), and phyla into domain (the highest rank or level). Groups of bacteria at each rank or level have names with endings or suffixes characteristic to that rank or level.

Table: Taxonomic ranks or levels in ascending order.

Rank or Level	Example
Species	E.coli
Genus	Escherichia
Family	Enterobacteriaceae
Order	Enterobacteriales
Class	γ-Proteobacteria
Phylum	Proteobacteria
Domain	Bacteria

Characteristics used in Bacterial Taxonomy

Classical Characteristics (Classical Taxonomy)

Several phenotypic characteristics (e.g., morphological, physiological and metabolic, ecological) and genetic analysis have been used in bacterial (microbial) taxonomy for many years.

These characteristics are assessed and the data are used to group bacteria up to the taxonomic ladder from species to domain. Classical characteristics are quite useful in routine identification of bacteria and also provide clues for phylogenic relationships amongst them as well as with other organisms.

Morphological Characteristics

Various morphological features, e.g., cell shape, cell size, colonial morphology, arrangement of flagella, cell motility mechanism, ultra structural characteristics, staining behaviour, endospore formation, spore morphology and location, and colour are normally used to classify and identify microorganisms.

Table: Some taxonomically useful morphological characteristics and their variations.

Characteristics	Variations
Cell morphology	
Unicellular	Cocci, bailliv, vibrios, spirilli, spirochaetes, prosthecate, stalked, sheathed.
Multicellular	Mycelial, filamentous.
Cell arrangement	Single, pairs, chains, bunches, packets.
Staining property	
Gram staining	Gram-positive, Gram-negative.
Acid fast staining	Acid-fast, non-acid fast.
Flagellation	Monotrichous, lophotrichous, amphritichous, peritrichous, endoflagellate or non-flagellate.
Motility	Non-motile, flagellar locomotion, gliding movement, motility due to endoflagella.
Glycocalyx	Capsule present or absent, slime layer.
Spores	Non-sporing, endospore, exospore, conidia, myxospores.
Sporangium	Shape, location of spore.
Cell inclusions	Poly β-hydroxybutyrate, volutin, polysaccharides, sulfur droplets, parasporal protein crystals.
Ultrastructural features	Surface structures of cells – flagella, pili, fimbriae, texture of slime layer.

Morphological characteristics play important role in microbial classification and identification due to following reasons:

- They are easily studied and analysed especially in eukaryotic microorganisms and comparatively complex prokaryotes.

- They normally do not vary greatly with environmental changes as they are resulted in by the expression of many genes and, therefore, are usually genetically stable.
- Morphological similarity amongst microorganisms often is a good indication of phylogenetic relatedness.

Some taxonomically useful morphological characteristics and their variations are shown in below table.

Physiological and Metabolic Characteristics

Some physiological and metabolic characteristics are very useful in classifying and identifying microorganisms because they are directly related to the nature and activity of microbial enzymes and transport proteins.

Some most important physiological and metabolic characteristics used in microbial taxonomy are nutritional types, cell wall components, carbon and nitrogen sources, energy metabolism, osmotic tolerance, oxygen relationships, temperature relationships, salt requirements and tolerance, secondary metabolites, storage inclusions, etc. Some taxonomically useful physiological and metabolic characteristics and their variations are given in table.

Table: Some taxonomically useful physiological and metabolic characteristics and their variations.

Character	Variations
Nutritional type	Photolithotrophs, chemolithotrophs, photorganotrophs, chemoorganotrophs.
Cell wall components	Peptidoglycans, teichoic and teichuronic acids, protein, polysaccharides, pseudomurein, etc.
Carbon sources	CO2+ sugars, sugar acids, sugar alcohols, polysaccharides, organic acids.
Nitrogen sources	Molecular nitrogen, ammonium salts, nitrate, organic nitrogenous compounds.
Energy metabolism	Photosynthesis, respiration, fermentation, inorganic substrate oxidation, nitrate and sulphate oxidation.
Oxygen relationship	Aerobic, microaerophilic, facultatively anaerobic, obligately anaerobic.
Temperature relationships	Mesophilic, facultatively thermophilic, obligately thermophilic, hyperthermophilic, psychrophilic.

Ecological Characteristics

Ecological characteristics, i.e., the characteristics of relationship of microorganisms to their environment significantly contribute in microbial taxonomy. It is because even very closely related microorganisms may vary considerably with respect to their ecological characteristics. For convenience, microorganisms inhabiting freshwater, marine, terrestrial, and human body environments differ from one another and from those living in different environments.

However, some of the most important ecological characteristics used in microbial taxonomy are – life cycle patterns, the nature of symbiotic relationship, pathogenic nature, and variations in the requirements for temperature, pH, oxygen, and osmotic concentrations.

Genetic Analysis

Genetic analysis is mostly used in the classification of eukaryotic microorganisms because the species is defined in these organisms in terms of sexual reproduction which occurs in them. This analysis is sometimes employed in the classification of prokaryotic microorganisms particularly those that use the processes of conjugation and transformation for gene exchange.

For example, members of genus Escherichia may conjugate with the members of genera Shigella and Salmonella but not with those of genera Enterobacter and Proteus. This shows that the members of first three genera are more closely related to one another than to Enterobacter and Proteus.

Studies of transformation with genera like Bacillus, Haemophilus Micrococcus, Rhizobium, etc. reveal that transformation takes place between different bacterial species but only rarely between genera. This provides evidence of a close relationship between species since transformation tails to occur unless the genomes are very much similar. Bacterial plasmids that carry genes coding tor phenotypic traits undoubtedly contribute in microbial taxonomy as they occur in most genera.

Molecular Characteristics

Some recent molecular approaches such as genomic DNA GC ratios, nucleic acid hybridization, nucleic acid sequencing, ribotyping, and comparison of proteins have become increasingly important and are used routinely for determining the characteristics of microorganisms to be used in microbial taxonomy.

Genomic DNA GC Ratios (G + C Content)

Genomic DNA GC ratio (G + C) is the first, and possibly the simplest, molecular approach to be used in microbial taxonomy. The GC ratio is defined as the percentage of guanine plus cytosine in an organism's DNA.

The GC ratio the base sequence and varies with sequence changes as follows:

$$\text{Mol}\%\text{G}+\text{C}=\frac{\text{G}+\text{C}}{\text{G}+\text{C}+\text{A}+\text{T}}\times 100$$

The GC ratio of DNA from animals and higher plants averages around 40% and ranges between 30 and 50%. Contrary to it, the GC ratio of both eukaryotic and prokaryotic microorganisms varies greatly; prokaryotic GC ratio is the most variable, ranging from around 20% to almost 80%. Despite such a wide range of variation, the GC ratio of strains within a particular species is constant.

Genomic DNA GC ratios of a wide variety of microorganisms have been determined, and knowledge of this ratio can be useful in microbial taxonomy, depending on the situation. For convenience, two microorganisms can possess identical GC ratios and yet turn out to be quite unrelated both taxonomically and phylogenetically because a variety of base sequences is possible with DNA of a single base composition.

In this case, the identical GC ratios are of no use with view point of microbial taxonomy. In contrast, if two microorganisms' GC ratio differs by greater than about 10%, they will share few DNA sequences in common and are therefore unlikely to be closely related.

GC ratio data are valuable in microbial taxonomy in at least two following ways:

- They can confirm a scheme of classification of microorganisms developed using other data. If microorganisms in the same taxon vary greatly in their GC ratios, the taxon deserves to be divided.
- GC ratio appears to be helpful in characterizing bacterial genera since the variation within a genus is usually less than 10% even though the content may vary greatly between genera.

For convenience, Staphylococcus and Micrococcus are the genera of gram-positive cocci having many features in common but differing in their GC ratio by more than 10%. The former has a GC ratio of 30-38%, whereas the latter of 64-75%.

Nucleic Acid Hybridization (Genomic Hybridization)

Nucleic acid hybridization or genomic hybridization measures the degree of similarity between two genomes (nucleic acids) and is useful for differentiating two bacteria (microorganisms). DNA-DNA hybridization is useful to study only closely related bacteria, whereas DNA-RNA hybridization helps comparing distantly related bacteria.

DNA-DNA Hybridization

Double-stranded DNA isolated from one bacterium is dissociated into single strands at appropriate temperature which are made radioactive with 32P, 3H, or 14C. Similarly, double-stranded DNA isolated from other bacterium is dissociated into single strands which are not made radioactive. The non-radioactive single-stranded DNA molecules are first allowed to bind to a nitrocellulose filter and unbound strands are removed by washing.

Now the filter with bound strands of DNA is incubated with the radioactive single-stranded DNA under optimal conditions of annealing. Annealing is an interesting feature of single-stranded DNA in which the strands, on cooling, tend to re-associate to form double-helix structure automatically.

Annealing occurs optimally when the temperature is brought to about 25 °C below the melting temperature (Tm) in a solution of high ionic concentration. However, during incubation the radioactive single strands of DNA hybridize with non-radioactive single strands of DNA depending on their homology in the base sequence. Then the filler is washed to remove unbound radioactive DNA molecules and the radioactivity of the hybridized radioactive DNA molecules is measured.

Following this, the amount of radioactivity in the hybridized radioactive DNA molecules is compared with the control which is taken as 100%, and this comparison gives a quantitative measure of the degree of complementarity of the two species of DNA, i.e., homology between the two DNAs.

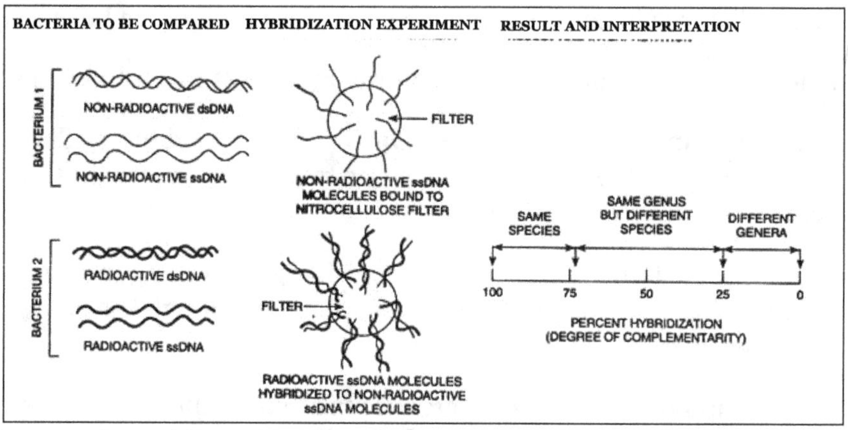

Figure: A schematic representation of DNA-DNA hybridization as a taxonomic tool.

Though there is no fixed convention as to how much hybridization between two DNAs is necessary to assign two bacteria to the same taxonomic rank, 70% or greater degree of complementarity of the two DNAs is recommended for considering the two bacteria belonging to the same species. By contrast, degree of at least 25% is required to argue that the two bacteria should reside in the same genus. Degree of complementarity to the range of 10% or less denotes that the two bacteria are more distantly related taxonomically.

DNA-DNA hybridization is a sensitive method for revealing subtle differences in the genes of two bacteria (other microbes also) and is therefore useful for differentiating closely related bacteria.

DNA homology studies have been conducted on more than 10,000 bacteria belonging to about 2,000 species and several hundred genera. It has proved to be a powerful tool in solving many problems of bacterial taxonomy, particularly at species level.

DNA-RNA Hybridization

DNA-RNA hybridization helps compare, unlike DNA-DNA hybridization, distantly related bacteria (microorganisms) using radioactive ribosomal RNA (rRNA) or transfer RNA (tRNA). It becomes possible because the DNA segments (genes) transcribing rRNA and tRNA represent only a small portion of the total DNA genome and have not evolved as rapidly as most other genes encoding proteins (i.e., they are more conserved in comparison to genes encoding proteins).

Among the different rRNAs, the 16S rRNA of prokaryotes and the analogous 18S rRNA of eukaryotic organisms have been found to be most suitable for comparison of their sequences is taxonomic studies. One of the major impacts of rRNA studies on taxonomy, for convenience, is the recognition of three major domains—the Archaea, the Bacteria, and the Eukarya by Woese and Colleagues in 1990.

DNA-RNA hybridization technique is similar to that employed for DNA-DNA hybridization. The filter-bound non-radioactive ssDNA is incubated with radioactive rRNA, washed, and counted. An even more accurate measurement of complementarity is obtained by finding the temperature required to dissociate and remove half the radioactive rRNA from the filter; the higher this temperature, the stronger the DNA-rRNA complex and the more similar the base sequences.

However, DNA-RNA hybridization has been done with thousands of bacteria for relevation of their taxonomic relationships. Such studies were made with pure cultures of bacteria till 1997-98, but since then, techniques have been developed to recover rRNA genes directly from natural habitats. This has come to be called as community analysis of rRNA from natural bacterial community.

Nucleic Acid Sequencing

Nucleic acid (DNA and RNA) sequencing is another molecular characteristic that helps directly compare the genomic structures. Most attention has been given to the sequencing of 5S and 16S rRNAs isolated from the 50S and 30S subunits of 70S prokaryotic ribosome, respectively.

rRNAs are almost ideal for the studies of bacterial (microbial) evolution and relatedness because:

- They are essential to ribosomes found in all bacteria.
- Their functions are same in all ribosomes.
- Their structure changes very slowly with time, i.e., they are more conserved.

The procedure of rRNA sequencing involves the following steps:

- rRNA is isolated from the ribosome and purified.
- Reverse transcriptase enzyme is used to make complementary DNA (cDNA) using primers that are complementary to conserved rRNA sequences.
- The cDNA is amplified using polymerase chain reaction (PCR) and finally.
- The cDNA is sequenced and the rRNA sequence deduced from the results.

Shotgun sequencing and other genome sequencing techniques have led to the characterization of many prokaryotic genomes (approximately 100) in a very short time and many more are in the process of being sequenced. Direct comparison of complete genome sequences undoubtedly will become an important tool in determining the classificatory categories of prokaryotes.

Ribotyping

Ribotyping is a technique which measures the unique pattern that is generated when DNA from a bacterium (all other organisms also) is digested by a restriction enzyme and the fragments are separated and probed with a rRNA probe.

The technique does not involve nucleic acid sequencing. Ribotyping has proven useful for bacterial identification and has found many applications in clinical diagnostics and for the microbial analyses of food, water, and beverages.

Ribotyping is a rapid and specific method for bacterial identification; it is so specific that it has been given the nickname 'molecular fingerprinting' because a unique series of bands appears for virtually any bacterium (any organism).

In ribotyping, first the DNA is isolated from a colony or liquid culture of the bacterium to be identified. Using polymerase chain reaction (PCR), genes of DNA for rRNA (preferably 16S rRNA) and related molecules are amplified, treated with one or more restriction enzymes, separated by electrophoresis, and then probed with rRNA genes.

The pattern generated from the fragments of DNA on the gel is then digitized and a computer used to make comparison of this pattern with patterns from other bacteria available from a database.

Comparison of Proteins

The amino acid sequences of proteins are direct reflections of mRNA sequences and therefore closely related to the structures of the genes coding for their synthesis. In the light of this, the comparisons of proteins from different bacteria prove very useful taxonomically.

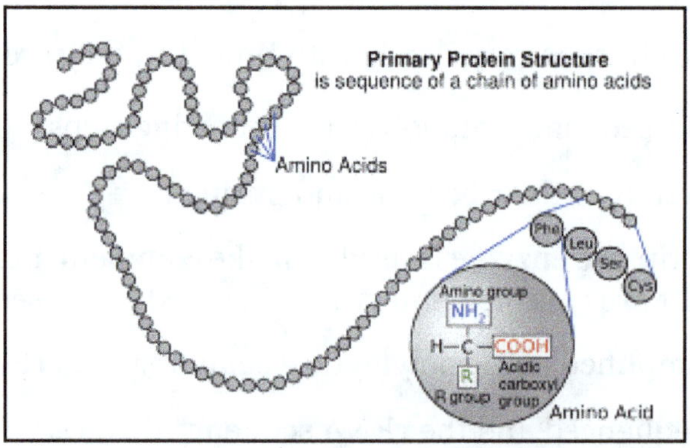

Although there are many methods to compare proteins, the most direct approach is to determine the amino acid sequences of proteins with the same function. When the sequences of proteins of the same functions in two bacteria are similar, the bacteria possessing them are considered to be closely related. However, the sequences of cytochromes and other electron transport proteins, histones, heat-sock proteins, and a variety of enzymes have been used in taxonomic studies.

References

- Microbiology: microscopemaster.com, Retrieved 12, July 2020
- What-are-the-branches-of-microbiology: ifas.ufl.edu, Retrieved 08, February 2020
- Bacteriology: microbenotes.com, Retrieved 23, August 2020
- Bacterial-taxonomy-meaning-importance-and-levels-54679: biologydiscussion.com, Retrieved 04, March 2020

Chapter 2

Bacteria and its Types

Bacteria are classified according to the nature of their cell walls, shapes or genetic makeup. Another way used to classify bacteria is Gram staining which further segregates them into Gram positive and Gram negative bacteria. This chapter explores the different types of bacteria to provide an extensive understanding of the subject.

Bacteria are single-celled microorganisms with prokaryotic cells, which are single cells that do not have organelles or a true nucleus and are less complex than eukaryotic cells. Bacteria with a capital B refers to the domain Bacteria, one of the three domains of life. The other two domains of life are Archaea, members of which are also single-celled organisms with prokaryotic cells, and Eukaryota. Bacteria are extremely numerous, and the total biomass of bacteria on Earth is more than all plants and animals combined.

Bacteria first arose on Earth approximately 4 billion years ago, and they were the first forms of life on Earth. For 3 billion years, bacteria and archaea were the most prevalent kinds of organisms on Earth. Multicellular eukaryotes did not appear until around 1.6-2 billion years ago. Eukaryotic cells, which make up all protists, fungi, animals, and plants, also contain what was once bacteria; it is thought that the mitochondria in eukaryotes, which produce energy through cellular respiration, and chloroplasts in plants and algae, which produce energy through photosynthesis, both evolved from bacteria that got taken up into cells in an endosymbiotic (mutually benefiting) relationship that became permanent over time.

Bacteria Characteristics

Bacteria are single-celled organisms. They lack organelles such as chloroplasts and mitochondria, and they do not have the true nucleus found in eukaryotic cells. Instead, their DNA, a double strand that is continuous and circular, is located in a nucleoid. The nucleoid is an irregularly shaped region that does not have a nuclear membrane. Bacteria also have a cell membrane and a cell wall that is often made of peptidoglycan. Together, the cell membrane and cell wall are referred to as the cell envelope. Many bacteria need a cell wall in order to survive.

Reproduction occurs through binary fission, which is the splitting of a bacterial cell after it reaches a certain size. Bacteria reproduce asexually, so the two daughter cells that result from binary fission have the same DNA as the parent cell. However, some bacteria can also exchange genetic material among one another in a process known as horizontal gene transfer. This method involves two already existing bacteria; it is not a form of transmission from parent to child.

Examples of Bacteria

Escherichia coli is one example of a common species of bacteria. It is bacillus-shaped and found naturally in the intestines of many animals including humans, where it produces vitamin K and b-complex vitamins. E. coli is also often used in laboratory research since it reproduces quickly and is hardy. Most strains of E. coli are harmless to humans, but some can cause infection. E. coli infection can result in gastrointestinal problems like diarrhea, and in more severe cases, bacterial meningitis or pneumonia can occur.

Lactobacillus acidophilus is another bacillus-shaped species of bacteria naturally found in places like the intestines and vagina, where it protects against harmful bacteria. It is a probiotic, a bacterium found in certain foods like yogurt and other fermented foods that are consumed in order to help absorb nutrients and replenish the body's supply of "good" bacteria. It can also be consumed in small amounts by people with lactose intolerance in order to help them consume lactose.

Some bacteria can be extremely harmful, such as Clostridium botulinum, the bacteria that causes botulism. C. botulinum produces the neurotoxin botulinum, which is responsible for the symptoms of botulism. Symptoms include blurred vision, nausea, trouble breathing, muscle weakness, and paralysis. Botulinum toxin is the deadliest known toxin; just one kilogram of botulinum would be deadly enough to kill the entire human population.

Sizes, Shapes and Arrangements of Bacteria

Bacterial Size

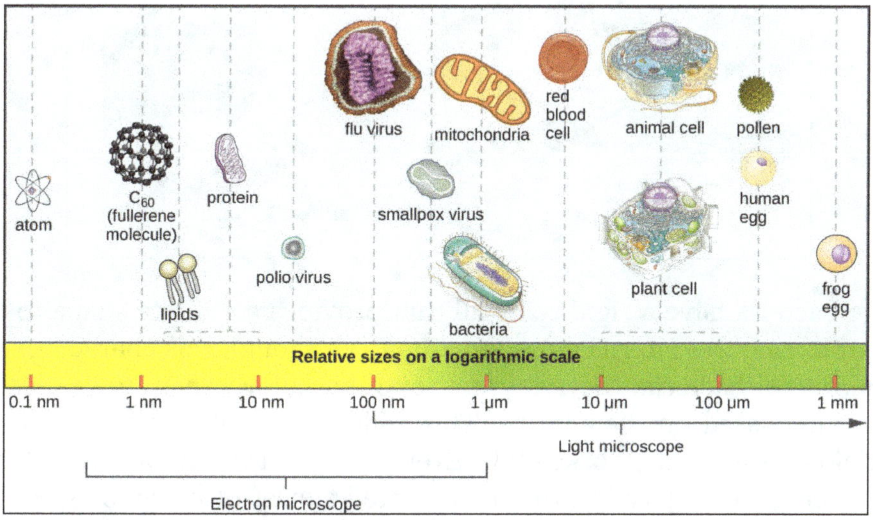

The relative sizes of various microscopic and nonmicroscopic objects. Note that a typical virus measures about 100 nm, 10 times smaller than a typical bacterium (~1 µm), which is at least 10 times smaller than a typical plant or animal cell (~10–100 µm). An object must measure about 100 µm to be visible without a microscope.

The unit of measurement used in bacteriology is the micron (micrometer) which is one-thousandth of a millimeter. Bacteria are, in general one-tenth the size of the eukaryotic cell. On average, the size of bacteria ranges from 0.5 to 5 µm. However, they can be as tiny as 0.3 µm and as large as 0.7 mm. The limit of resolution with the unaided eye is about 200 microns, and as many bacteria are smaller than this size, they are not visible with naked eyes. Among the largest bacteria is Thiomargarita namibiensis, which is up to half a millimeter long and Epulopiscium fishelsoni which is 0.75 mm long. The smallest bacteria are members of genus Mycoplasma which are only 0.3 µm, as small as the largest viruses. The size of common bacteria like Escherichia coli ranges in size from 1.1 to 1.5 µm in diameter. It has been observed that the size of bacteria has a significant role in the survival of the organisms. Owing to their tiny size, they are capable of surviving and even thriving in various unlikely environments like the vertical sediments in the marine environment.

Since other organisms are absent in such an environment, bacteria can utilize the available resources. Besides, the small size of bacteria favors parasitism and the ability to survive in areas with low nutrition. The high surface area-volume ratio also allows the bacteria to take up all the nutrients required for survival while allowing the steady growth and reproduction.

Bacterial Shape

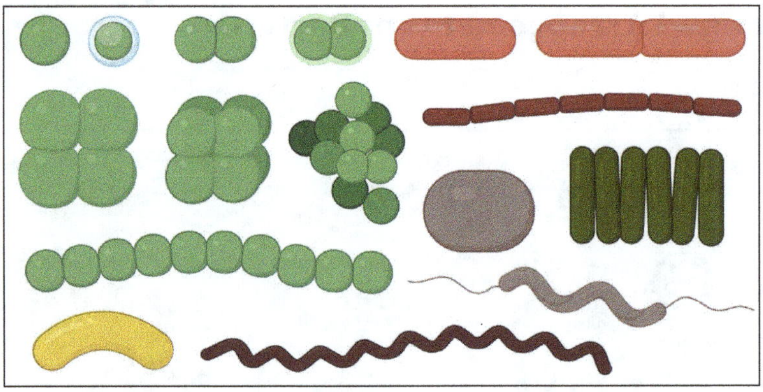

Most of the bacteria have a rigid cell wall that provides a definite shape to the bacteria while protecting the internal components. Even though this characteristic is valid for the majority of bacteria, they vary in shape that allows them to be classified into different groups based on their forms. This wide variety of shapes is determined by the bacterial cell wall and cytoskeleton. Even though bacteria have a wide variety of shapes, anyone genus typically exhibits a limited subset of morphologies, indicating that, with a universe of shapes to choose from, individual bacteria adopt only those that are adaptive. Bacteria with different shapes present different physical features to the outside world, and these features help cells cope with and adapt to external conditions. It has been observed that bacterial shape contributes a measure of survival value in the face of nutrient acquisition, cell division, predators, attachment to surfaces, passive dispersal, active motility, and internal or external differentiation.

The common categories of bacteria based on their shapes are:

Cocci

The bacteria that are oval or spherical in shape are included called cocci bacteria. These may either remain single or attached to one another in groups. They appear flattened when placed in groups. It is assumed that coccoid forms were derived from rod-shaped organisms through evolutionary time.

Bacilli (Rod-shaped)

These are rod-shaped cells that also like cocci, remain either single or attached to other cells. Bacilli bacteria are among the first bacteria to have arisen, and this shape is said to be not as advantageous as other shapes. This has been assumed upon the observation of the behavior of filamentous E. coli cells which, though motile and chemotactic, move slowly and cannot tumble to change direction.

Spiral

This group includes bacteria that are either helical-shaped or curved (comma-shaped). The bacteria can range from slightly curved to corkscrew-like spiral.

Arrangements of Cocci

Cocci bacteria can be arranged either singly, in pairs, in groups of four, in chains, in clusters or cubes consisting of eight cells. These cells remain attached during cell division.

Coccus

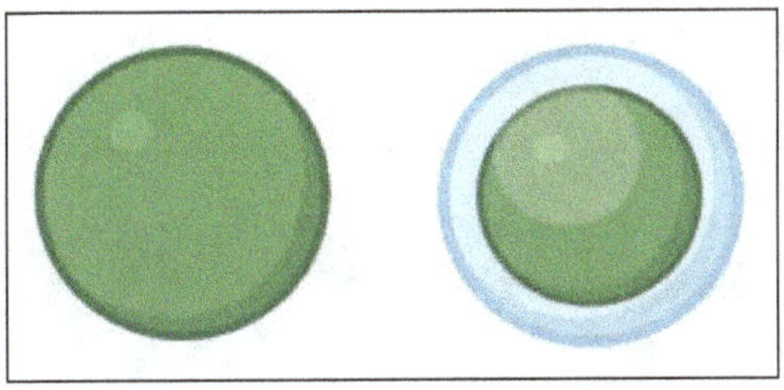

This group includes bacteria that are present as a single cell.

Diplococci

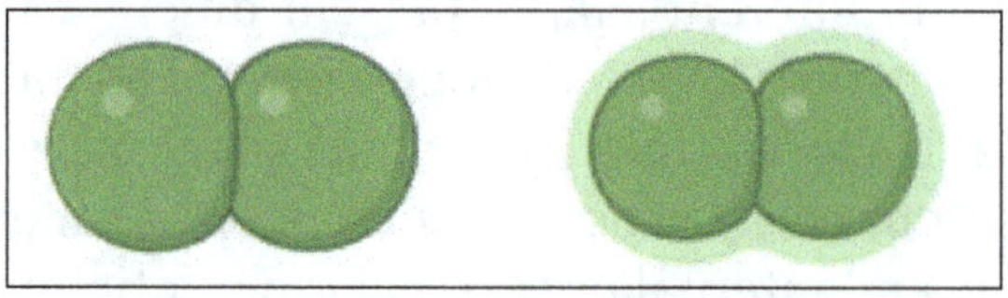

This arrangement results when two bacterial cells occur as a pair (joined together). Some of the cells in this arrangement might remain spherical while some might appear flattened, elongated, or bean-shaped. Examples: Streptococcus pneumonia, Moraxella catarrhalis, Enterococcus spp, Neisseria gonorrhea.

Tetrad

Tetrad bacteria are arranged in a group of four cells that remain attached and grow in the attachment after cell division. This arrangement results when the cells divide into two planes.Examples: Aerococcus, Pediococcus, and Tetragenococcus.

Sarcina

In this arrangement, the bacterial cells form a group of eight cells. This happens when the cells divide in a perpendicular plane. The common characteristic associated with these organisms is being strict anaerobe. Examples: Sarcina aurantiaca, Sarcina lutea, Sarcina ventriculi.

Streptococci

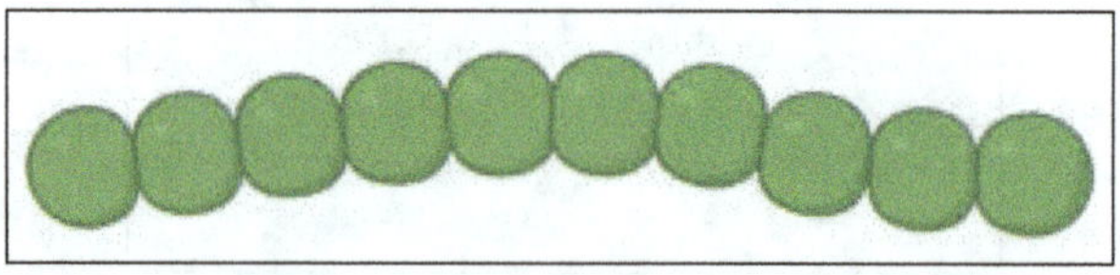

Here, the bacteria are arranged in long chains. These bacteria are present in family

Streptococcaceae, which is characterized by a lack of motility and Gram-positive bacteria. Examples: Streptococcus pyogenes, Streptococcus pneumonia, Streptococcus mutans.

Staphylococci

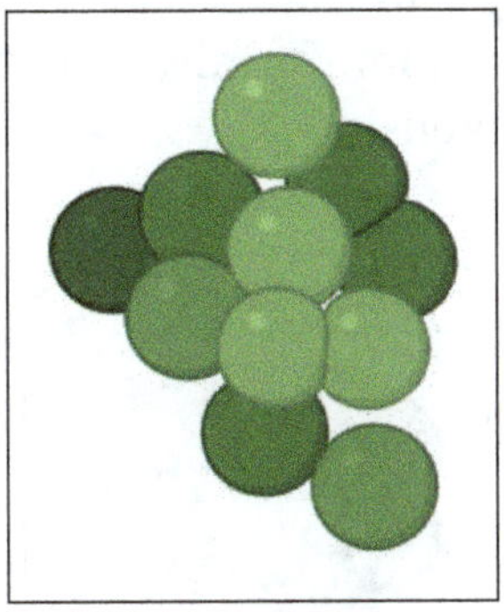

This type includes bacteria that are arranged in grape-like clusters. This results from cell division in both the planes and is characterized by organisms which are immotile and Gram-positive. Examples: Staphylococcus epidermidis, Staphylococcus haemolyticus, Staphylococcus aureus, Staphylococcus capitis.

Arrangement of Bacilli

Bacillus

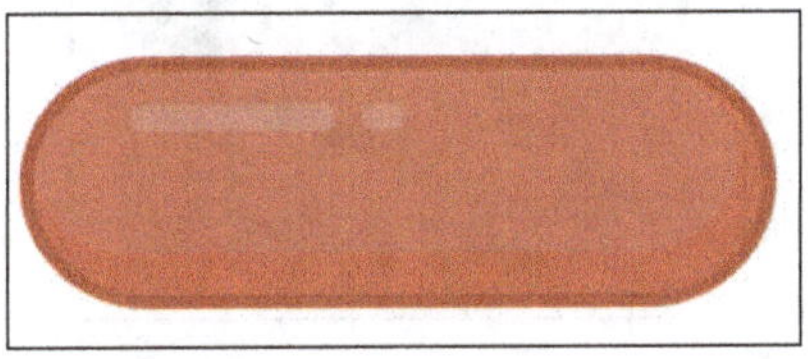

Bacilli are the bacteria which are rod-shaped and are present as single cells. These bacteria can form endospores and are facultative anaerobes. Examples: Salmonella enterica subsp, Bacillus cereus, and Salmonella choleraesuis.

Diplobacilli

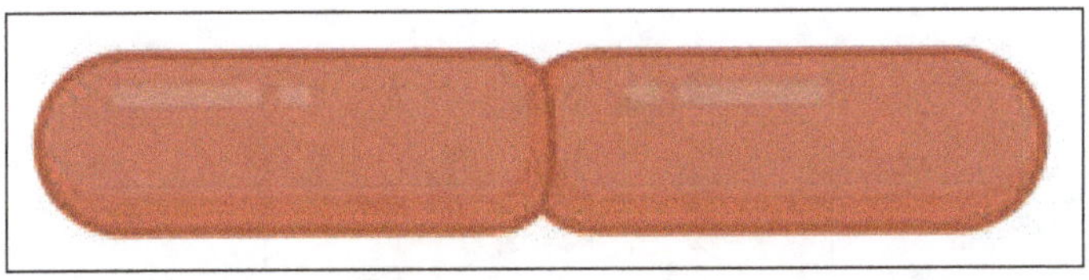

As in Diplococci, Diplobacilli also exists in pairs. After cell division, the two cells do not divide and grow in an attached arrangement. Examples: Coxiella burnetii, Klebsiella rhinoscleromatis, Moraxella bovis.

Streptobacillus

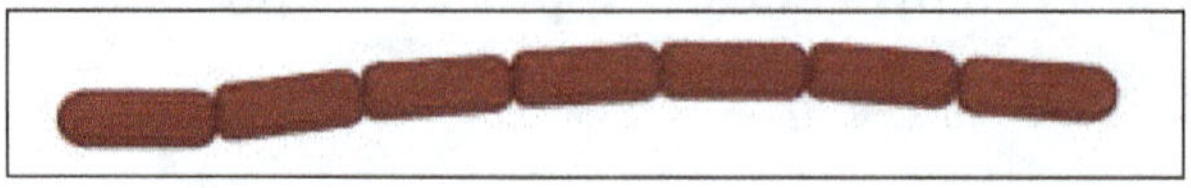

In this group, bacteria are arranged in chains. This results from cell division in a single chain. Examples: Streptobacillus moniliformis, Streptobacillus Levaditi, Streptobacillus felis, Streptobacillus hongkongensis.

Coccobacilli

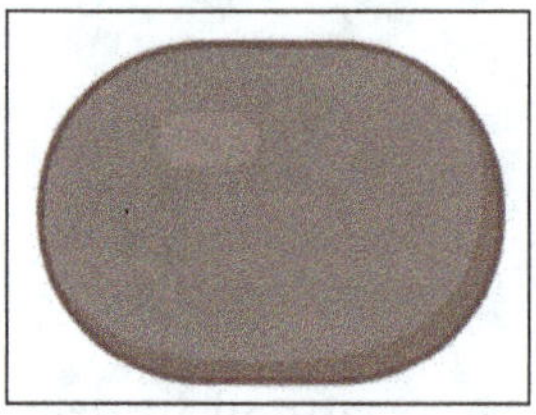

As the name suggests, coccobacilli resemble both cocci as well as bacilli. These are shorter in size and thus, appear stumpy. Examples: Chlamydia trachomatis, Haemophilus influenza, Gardnerella vaginalis.

Pallisades

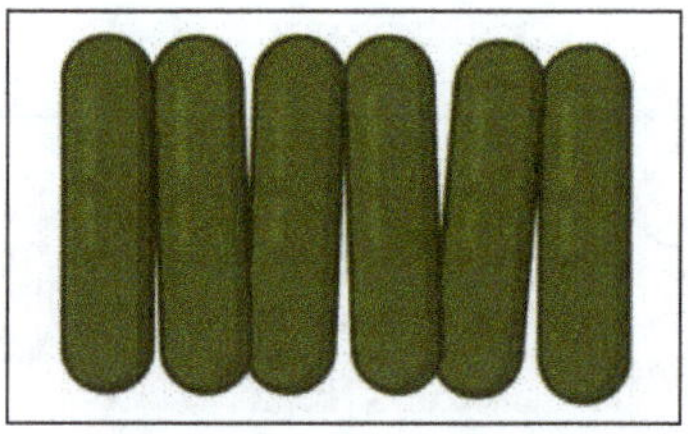

Pallisades are the type of bacilli bacteria that resemble a picket fence structure as a result of the bent at the point of division during cell division. They appear similar to Chinese letters. Example: Corynebacterium diphtheria that causes diphtheria.

Arrangement of Spiral

Vibrio

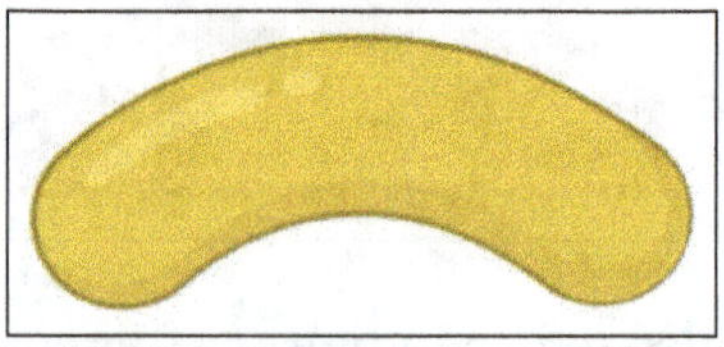

These are the slightly curved bacteria resembling a comma shape. Examples: Vibrio mytili, Vibrio anguillarum, Vibrio parahaemolyticus, Vibrio cholera.

Spirochetes

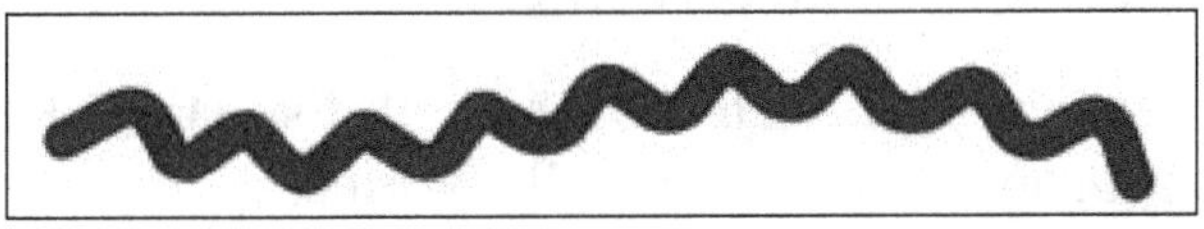

Spirochetes are spiral bacteria having a helical shape. These are flexible and have an axial filament which helps in motility. These filaments are essential distinguishing character between spirochetes and other bacteria. These filaments run throughout the length of the bacteria and thus, help in twisting the motion of the bacteria. Examples: Leptospiraspecies (Leptospira interrogans), Treponema pallidum, Borrelia recurrentis.

Spirilla (Helical-shape/Corkscrew Form)

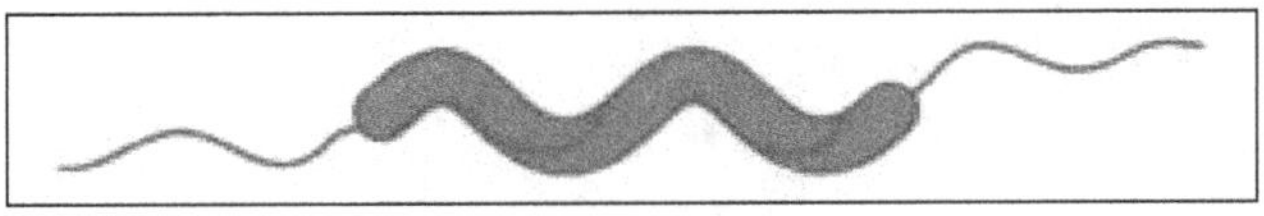

These bacteria are similar in structure with spirochetes but are more rigid. They, too, have a flagellum but lack the endoflagella like in spirochetes. Examples: Campylobacter jejuni, Helicobacter pylori, Spirillum winogradskyi.

Other Shapes and Arrangements

- Appendaged Bacteria: The bacteria that produce a unique structure like pillus or fimbriae are called appendaged bacteria. These bacteria are more virulent than other bacteria that do not form these appendages. Example: Neisseria gonorrheae, the agent of Gonorrhea.
- Box-shaped/Rectangular Bacteria: Box-shaped bacteria are rectangular in shape and resemble a box. Example: Haloarcula marismortui.
- Club-shaped Rod Bacteria: These bacteria are thinner on one side than the other. One of the classic examples of this group is Corynebacterium.
- Filamentous Bacteria: These are bacteria that are long, thin, and filament-shaped. They, sometimes, divide to form branches resembling strands of hair or spaghetti called mycelium. Example: Actinomycetes.
- Triangular-shaped Bacteria: This group includes bacteria that are triangular in shape. Example: Haloarcula.
- Pleomorphic Bacteria: The bacteria that do not have a specified shape are included in this group. They can change their shape, but in pure culture, they appear to have a definite form. Examples: Mycoplasma pneumoniae, M. genitalium.

- Stalked Bacteria: These are the bacteria that possess a stalk on one end of the cell. Examples: Caulobacter crescentus.
- Star-shaped Bacteria The bacteria that look like stars or are star-shaped are included in this group. Examples: Stella humosa.

Bacterial Structure

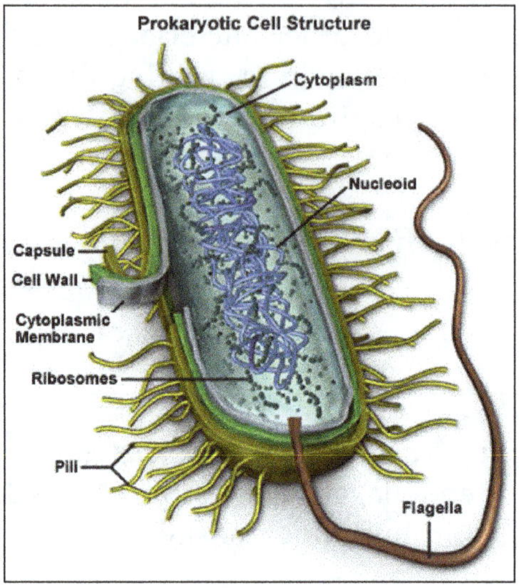

Bacteria are prokaryotes, lacking well-defined nuclei and membrane-bound organelles, and with chromosomes composed of a single closed DNA circle. They come in many shapes and sizes, from minute spheres, cylinders and spiral threads, to flagellated rods, and filamentous chains. They are found practically everywhere on Earth and live in some of the most unusual and seemingly inhospitable places.

Evidence shows that bacteria were in existence as long as 3.5 billion years ago, making them one of the oldest living organisms on the Earth. Even older than the bacteria are the archeans (also called archaebacteria) tiny prokaryotic organisms that live only in extreme environments: boiling water, super-salty pools, sulfur-spewing volcanic vents, acidic water, and deep in the Antarctic ice. Many scientists now believe that the archaea and bacteria developed separately from a common ancestor nearly four billion years ago. Millions of years later, the ancestors of today's eukaryotes split off from the archaea. Despite the superficial resemblance to bacteria, biochemically and genetically, the archea are as different from bacteria as bacteria are from humans.

In the late 1600s, Antoni van Leeuwenhoek became the first to study bacteria under the microscope. During the nineteenth century, the French scientist Louis Pasteur and the German physician Robert Koch demonstrated the role of bacteria as pathogens (causing disease). The twentieth century saw numerous advances in bacteriology, indicating their diversity, ancient lineage, and general importance. Most notably, a number of scientists around the world made contributions to the field of microbial ecology, showing that bacteria were essential to food webs and for the overall health of the Earth's ecosystems. The discovery that some bacteria produced compounds lethal to other bacteria led to the development of antibiotics, which revolutionized the field of medicine.

There are two different ways of grouping bacteria. They can be divided into three types based on their response to gaseous oxygen. Aerobic bacteria require oxygen for their health and existence and will die without it. Anerobic bacteria can't tolerate gaseous oxygen at all and die when exposed to it. Facultative aneraobes prefer oxygen, but can live without it.

The second way of grouping them is by how they obtain their energy. Bacteria that have to consume and break down complex organic compounds are heterotrophs. This includes species that are found in decaying material as well as those that utilize fermentation or respiration. Bacteria that create their own energy, fueled by light or through chemical reactions, are autotrophs.

- Capsule - Some species of bacteria have a third protective covering, a capsule made up of polysaccharides (complex carbohydrates). Capsules play a number of roles, but the most important are to keep the bacterium from drying out and to protect it from phagocytosis (engulfing) by larger microorganisms. The capsule is a major virulence factor in the major disease-causing bacteria, such as Escherichia coli and Streptococcus pneumoniae. Nonencapsulated mutants of these organisms are avirulent, i.e. they don't cause disease.

- Cell Envelope - The cell envelope is made up of two to three layers: the interior cytoplasmic membrane, the cell wall, and - in some species of bacteria - an outer capsule.

- Cell Wall - Each bacterium is enclosed by a rigid cell wall composed of peptidoglycan, a protein-sugar (polysaccharide) molecule. The wall gives the cell its shape and surrounds the cytoplasmic membrane, protecting it from the environment. It also helps to anchor appendages like the pili and flagella, which originate in the cytoplasm membrane and protrude through the wall to the outside. The strength of the wall is responsible for keeping the cell from bursting when there are large differences in osmotic pressure between the cytoplasm and the environment.

  Cell wall composition varies widely amongst bacteria and is one of the most important factors in bacterial species analysis and differentiation. For example, a relatively thick, meshlike structure that makes it possible to distinguish two basic types of bacteria. A technique devised by Danish physician Hans Christian Gram in 1884, uses a staining and washing technique to differentiate between the two forms. When exposed to a gram stain, gram-positive bacteria retain the purple color of the stain because the structure of their cell walls traps the dye. In gram-negative bacteria, the cell wall is thin and releases the dye readily when washed with an alcohol or acetone solution.

- Cytoplasm - The cytoplasm, or protoplasm, of bacterial cells is where the functions for cell growth, metabolism, and replication are carried out. It is a gel-like matrix composed of water, enzymes, nutrients, wastes, and gases and contains cell structures such as ribosomes, a chromosome, and plasmids. The cell envelope encases the cytoplasm and all its components. Unlike the eukaryotic (true) cells, bacteria do not have a membrane enclosed nucleus. The chromosome, a single, continuous strand of DNA, is localized, but not contained, in a region of the cell called the nucleoid. All the other cellular components are scattered throughout the cytoplasm.

  One of those components, plasmids, are small, extrachromosomal genetic structures carried by many strains of bacteria. Like the chromosome, plasmids are made of a circular piece of DNA. Unlike the chromosome, they are not involved in reproduction. Only the chromosome has the genetic instructions for initiating and carrying out cell division, or binary fission, the primary means of reproduction in bacteria. Plasmids replicate independently of the chromosome and, while not essential for survival, appear to give bacteria a selective advantage.

  Plasmids are passed on to other bacteria through two means. For most plasmid types, copies in the cytoplasm are passed on to daughter cells during binary fission. Other types of plasmids, however, form a tubelike structure at the surface called a pilus that passes copies of the plasmid to other bacteria during conjugation, a process by which bacteria exchange genetic information. Plasmids have been shown to be instrumental in the transmission of special properties, such as antibiotic drug resistance, resistance to heavy metals, and virulence factors necessary for infection of animal or plant hosts. The ability to insert specific genes into plasmids have made them extremely useful tools in the fields of molecular biology and genetics, specifically in the area of genetic engineering.

- Cytoplasmic Membrane - A layer of phospholipids and proteins, called the cytoplasmic membrane, encloses the interior of the bacterium, regulating the flow of materials in and out of the cell. This is a structural trait bacteria share with all other living cells; a barrier that allows them to selectively interact with their environment. Membranes are highly organized and asymmetric having two sides, each side with a different surface and different functions. Membranes are also dynamic, constantly adapting to different conditions.

- Flagella - Flagella (singular, flagellum) are hairlike structures that provide a means of locomotion for those bacteria that have them. They can be found at either or both ends of a bacterium or all over its surface. The flagella beat in a propeller-like motion to help the bacterium move toward nutrients; away from toxic chemicals; or, in the case of the photosynthetic cyanobacteria; toward the light.

- Nucleoid - The nucleoid is a region of cytoplasm where the chromosomal DNA is located. It is not a membrane bound nucleus, but simply an area of the cytoplasm where the strands of DNA are found. Most bacteria have a single, circular chromosome that is responsible for replication, although a few species do have two or more. Smaller circular auxiliary DNA strands, called plasmids, are also found in the cytoplasm.

- Pili - Many species of bacteria have pili (singular, pilus), small hairlike projections emerging from the outside cell surface. These outgrowths assist the bacteria in attaching to other cells and surfaces, such as teeth, intestines, and rocks. Without pili, many disease-causing bacteria lose their ability to infect because they're unable to attach to host tissue. Specialized pili are used for conjugation, during which two bacteria exchange fragments of plasmid DNA.
- Ribosomes - Ribosomes are microscopic "factories" found in all cells, including bacteria. They translate the genetic code from the molecular language of nucleic acid to that of amino acids—the building blocks of proteins. Proteins are the molecules that perform all the functions of cells and living organisms. Bacterial ribosomes are similar to those of eukaryotes, but are smaller and have a slightly different composition and molecular structure. Bacterial ribosomes are never bound to other organelles as they sometimes are (bound to the endoplasmic reticulum) in eukaryotes, but are free-standing structures distributed throughout the cytoplasm. There are sufficient differences between bacterial ribosomes and eukaryotic ribosomes that some antibiotics will inhibit the functioning of bacterial ribosomes, but not a eukaryote's, thus killing bacteria but not the eukaryotic organisms they are infecting.

Bacterial Spores

Bacterial species have different coping mechanisms with selective harsh environmental conditions. One of the most common coping mechanisms for bacteria is forming spores to protect themselves against ecological degrading agents. Bacterial spores are the most dormant form of bacteria since they exhibit minimal metabolism and respiration, as well as reduced enzyme production.

Typically, Gram-positive bacteria are best known for producing intracellular spores called endospores as a survival mechanism. Endospores are highly retractile and thick-walled structures formed inside the bacterial cells. It is most common for Bacillus species as well as Clostridium species to create endospores. B. cereus is a member of Bacillus species and is well-known for its ability to cause foodborne illness as a result of its spores surviving various temperatures. Similarly, C. perfringens spores are acid-soluble proteins that show high resistance to chemicals and heat.

Endospores can resist inactivation from ethanol treatment. They also can survive high temperatures for up to 150 °C, making specific Gram-positive species heat resistant. Further, bacterial spores can show typical viability signs at temperatures near the absolute zero. Endospores are resistant to the chemical agents e.g., triphenylmethane dyes and can even protect the bacterial cells against ultraviolet radiation, extreme pH gradients, drought, as well as nutrition depletion.

Endospores germinate back into vegetative cells (an active bacterial cell that undergoes metabolism) when surrounding environmental conditions favor bacterial growth and reproduction. Several stimulants revert bacterial cells to their active vegetative cells, such as optimal close-to-body temperature as well as diffusion of nutrients and water through bacterial cell walls through alteration of their surface tension.

The process of spore formation is a multistep process. It starts from replication of the bacterial DNA, followed by the formation of the forespore, which is, by definition, pinching of the cellular plasma membrane between the replicated chromosomes. Then, a cortex forms between the inner and outer membrane by extending the second cellular membrane to enclose forespore with calcium and dipicolinic acid. Finally, the external spore coat surrounds the endospore before its release.

Microscopic examination to delineate the morphology of endospores involves differential staining processes such as malachite green and fluorescence staining techniques. Staining dormant bacterial samples with malachite green as the primary stain and safranine as the counteract stain results in the appearance of green oval endospores enclosed inside pink vegetative bacterial cells. There are different locations of the endospores inside the bacterial cell. For instance, central endospores are located in the middle of the bacterial cell, while the terminal endospore appears at the end. There is also a subterminal type of endospores which appears between the middle and the end of the cell.

Despite their sturdy and resistant nature to environmental threats, endospores can get affected by certain eradication factors. During the 17th century, John Tyndall, a famous European physicist, discovered Tyndallization. The latter is the process of heating liquids and objects at a temperature of 80 to 100 °C for 30 minutes; then, the sample is incubated. The procedure is repeated for three consecutive days. The principle behind successive heating for three days is that heating endospores for the first-time results in reverting them into vegetative cells killed through repetitive heat in the second and third days.

Anthrax Bioweapon (Bioterrorism)

B. anthracis is a gram-positive spore-forming bacteria that are commonly found in the soil of endemic areas. It is one of the most common agents used in biological warfare. Many factors make *B. anthracis* as a good bioweapon. Its endospores can be placed into food, water, powder, and sprays spreading the anthrax infection without anyone's knowledge, as endospores are microscopic, cannot be tasted or felt. In 2001, anthrax spores were used as a bioweapon in the U.S. They were distributed into letters delivered by the United States Postal system spreading anthrax infection among 22 mail handlers and customers. Anthrax bioterrorism attacks can take several other forms. B. anthracis endospores can be released into food and water, sprayed from the air or high buildings, or even carried on clothes or shoes. B. anthracis endospores can cast a high risk of misuse and pose severe threats to public safety and health.

Clostridium Difficile Colitis

C. difficle infections (CDI) are associated with high morbidity, mortality and healthcare costs. The cost of CDI is estimated at $5.4 billion in the United States - $4.7 billion (86.7%) from healthcare settings and $725 million (13.3%) incurred in the community. The morbidity includes an increased need for colectomies, discharges to nursing homes and readmissions. The morbidity and mortality rates of pseudomembranous colitis are 10% to 20% in untreated elderly individuals. In patients with toxic megacolon, the mortality can be as high as 35% despite surgical intervention.

Clinical Significance

Anthrax

Endospores of B. anthracis cause anthrax. It has four types according to its mode of infection and the affected system:

- Cutaneous anthrax,
- Gastrointestinal anthrax,
- Inhalational anthrax,
- Injection anthrax.

Cutaneous anthrax develops as a result of wound contamination with B. anthracis endospores while handling contaminated animal products. It is characterized by a painless skin ulcer with a black center and edematous blisters that may be itchy. Gastrointestinal anthrax develops as a result of the ingestion of water or food contaminated with B. anthracis endospores. The patients can present with fever, nausea, bloody vomiting, bloody diarrhea, painful swellings of the neck lymph nodes, and flushing of the face and the eyes. Inhalational anthrax is the result of the inhalation of bacterial endospores while handling contaminated animal materials, such as wools or feces. It characteristically presents with fever, shortness of breath, cough, chest discomfort, body ache, sweating, and nausea. Injection anthrax develops from using syringes contaminated with bacterial endospores. It clinically presents with the development of deep abscess under the skin.

Appropriate specimen collection is crucial to confirm the diagnosis. For inhalational anthrax, gram stain and PCR of pleural, bronchial, and CSF (if meningitis) is usually required. Skin swabs for gram stain, PCR, and biopsy of the skin lesion are performed in cutaneous anthrax. Similarly, gram stain, culture, and PCR of ascitic fluid, oral lesion, or rectal swab can be collected in gastrointestinal anthrax. Systemic or disseminated anthrax is treated with a combination of intravenous antibiotics (ciprofloxacin, meropenem, and Linezolid) and antitoxin therapy.

Tetanus

It is a bacterial infection caused by contamination of open wounds with C. tetani endospores. The most common lesions owing to tetanus infection are contaminated puncture wounds, infected foot ulcers, surgical wounds, and animal bites. Within the incubation period of tetanus (7 to 10 days), the bacterial endospores produce a neurotoxin called tetanospasmin that impaired motor nerves causing the characteristic clinical picture of tetany. The most common symptoms of tetanus are trismus (stiffness of the jaw muscles), painful body spasms especially in the neck and abdominal muscles triggered by a noise or physical touch in addition to difficulties in swallowing and fever. Tetanic contractions strengthen through the course of infection and may cause fractures and pulmonary embolism, which eventually leads to death. The diagnosis of tetanus can be through toxin assays in the blood. Tetanus has no treatment. However, prevention of tetanus is through vaccination with the DTap vaccine given as a series of injections during childhood (with the first delivered at two months of age while the last vaccination given at the age of 4). Also, as further prevention from the infection, booster doses against the toxin are given once every ten years.

Food Poisoning

Endospores causing food poisoning include:

- Bacillus cereus,
- Clostridium perfringes.

B. cereus endospores are among the leading organisms causing food poisoning. B. cereus food poisoning is divided into emetic and diarrheal subtypes and caused mainly by the ingestion of raw and contaminated food with B. cereus endospores. Emetic food poisoning results from B. cereus endospores that produce cereulide toxin, which nausea and vomiting. Diarrheal food poisoning results from the ingestion of meat products, milk, or vegetable contaminated with enterotoxin secreting endospores. Diarrheal syndrome characteristically presents with bloody or mucoid diarrhea and abdominal pain.

C. perfringens is naturally present in the intestinal microbiota. Ingestion of food contaminated with human or animal feces containing C. perfringens endospores causes' food poisoning. The symptoms usually appear 6 to 24 hours following the ingestion of contaminated food and characterized by abdominal cramps and watery diarrhea. The diagnosis is through stool analysis and culturing bacteria on differential microbiological media. Diagnosis of both bacterial strains can be confirmed through isolation of bacteria from feces or vomitus and culturing it on a differential media plate. Patients usually recover without the administration of antibiotics. Supportive treatment with fluid administration is the key due to excessive diarrhea and vomiting.

Clostridial Myonecrosis

Clostridial myonecrosis (gas gangrene) is a bacterial infection caused by clostridial endospores (especially C. perfringens) and most commonly affects upper and lower extremities. There are two types of gas gangrene: Traumatic and Non-traumatic syndromes. The traumatic syndrome is the most common and includes contamination of open wounds with Clostridium endospores. In contrast, the non-traumatic syndrome develops diminished perfusion from vascular diseases (such as atherosclerosis) or diabetes mellitus leads to clostridial spores infection and endotoxin mediated gas gangrene. The common symptoms of gas gangrene are blisters with a foul smell, painful edema around the wound, air under the skin, fever, and jaundice (yellow skin and eyes) at later stages. The diagnosis involves skin culture to test for C. perfringens. Surgical evaluation is necessary to limit the spread of gangrene. The treatment rests on prompt administration of antibiotics and surgical removal of necrotic tissues. Late-stage may require limb amputation.

Clostridium Difficile Colitis

C. difficile is part of the healthy intestinal microbiota. However, overgrowth of the bacteria as a result of prolonged intake of antibiotics (such as fluoroquinolones, clindamycin, and penicillin) disrupts the balance of colon microbiota, causing pseudomembranous colitis. This syndrome is commonly considered a nosocomial infection. C. difficile colitis contributes to about 15% - 30% of antibiotic-associated diarrhea. Common symptoms of this condition include fever, watery diarrhea, nausea, mucous in stool, and abdominal cramps. About 8% of C. difficile infections develop a fulminant infection.

Botulism

C. botulinum endospores cause four significant types of botulism syndrome according to the mode of infection. Foodborne botulism is caused by contamination of food with bacterial endotoxin and spores. Wound botulism is caused by the toxin produced by the colonization of C.botulinum inside open wounds. Intestinal colonization of C. botulinum in infants and adults (though it is rare) causing infant botulism. Inhalation botulism is caused by inhaling C. botulinum toxins. Besides, there is another rare form of botulism called iatrogenic botulism that results from the injection of C. botulinum toxins such as rarely failed procedures of botox injection during cosmetic surgeries.

The clinical picture of all types of botulism consists of the symptoms starting from symmetrical cranial nerve paralysis, followed by descending symmetrical flaccid paralysis of voluntary muscles. Those symptoms progress to respiratory muscle paralysis leading to respiratory failure and death. The exact dose of the lethality of botulism toxins is not accurately defined. However, the commonly estimated lethal dose of pure type A botulinum crystalline is about 0.1 micrograms for a 70-kg man.

The mortality rates attributed to botulism, have improved since 1910 due to the development of intensive care techniques, such as mechanical ventilation. Patients with suspected botulism should be promptly referred to intensive care units closer monitoring in case of progressive respiratory failure. Paralysis caused by botulinum toxin is prolonged and could last for several weeks or months. Rehabilitation of paralyzed patients is crucial.

The treatment of botulism is antitoxin therapy. Antitoxin therapy is usually given during the first 24 hours after the exhibition of symptoms. Botulinum toxin can be absorbed through mucus membranes of the eye, mouth, and nose even though they are not transmitted through direct skin contact. Therefore proper isolation and PPE (personal protective equipment) are essential.

Bacterial Flagella

Bacterial flagella are long, thin (about 20 nm), whip-like appendages that move the bacteria towards nutrients and other attractants. Flagella are free at one end and attached to the cell at the other end. Flagellum can never be seen directly with the light microscope but only after staining with special flagella stains that increase their diameter.

Flagella are usually found in gram-negative bacilli. Gram-positive rods (e.g., Listeria species) and cocci (some Enterococcus species, Vagococcus species) also have flagella. The long filament of flagella is composed of many subunits of a single protein, flagellin, arranged in several intertwined chains. The energy for movement, the proton motive force, is provided by ATP.

Flagella are helical shaped structure composed of subunits of a protein called flagellin. The wider region at the base of the flagellum is called a hook. It is different in structure than that of the filament. Hook connects filament to the motor portion of the flagellum called a basal body.

The basal body is anchored in the cytoplasmic membrane and cell wall. There are presences of rings that are surrounded by a pair of proteins called Mot. These proteins actually drive the flagellar motor causing rotation of the filament. Another set of proteins called Fli proteins function as the motor switch, reversing the rotation of the flagella in response to intracellular signals.

Arrangement and Types of Bacterial Flagella

The number and location of flagella are distinctive for each genus. There are four types of flagellar arrangement:

- Monotrichous (Mono means one): Single polar flagellum e.g. Vibrio cholerae, Campylobacter spp. (polar flagella often in pairs to give a "seagull" appearance).
- Amphitrichous: Single flagellum at both ends e.g. Alcaligenes faecalis (Note: amphibians live both on land and in water).
- Lophotrichous: Tuft of flagella at one or both ends e.g. Spirilla spp.

- Peritrichous (flagella in periphery): Flagella surrounding the bacterial cell. All the members of family Enterobacteriaceae, if motile have peritrichous flagella. e.g. Salmonella Typhi, Escherichia coli, Proteus spp (highly motile organism; shows swarming motility).

Functions of Bacterial Flagella

Many prokaryotes are motile, and the majority of motile prokaryotes move by means of flagella.

Medical Importance of Flagella

- Role in Pathogenesis: Escherichia coli and Proteus spp are common causes of urinary tract infections. The flagella of these bacteria help the bacteria by propelling up the urethra into the bladder.
- Roles in Organism identification:
    - Some species of bacteria, eg. Salmonella species are identified in the clinical laboratory by the use of specific antibodies against flagellar proteins.
    - Organisms such as Vibrio cholerae (darting motility) and Proteus species (swarming growth in common culture media) are easily identified by their characteristics motility pattern.

Bacterial Metabolism

Metabolism refers to all the biochemical reactions that occur in a cell or organism. The study of bacterial metabolism focuses on the chemical diversity of substrate oxidations and dissimilation reactions (reactions by which substrate molecules are broken down), which normally function in bacteria to generate energy. Also within the scope of bacterial metabolism is the study of the uptake and utilization of the inorganic or organic compounds required for growth and maintenance of a cellular steady state (assimilation reactions). These respective exergonic (energy-yielding) and endergonic (energy-requiring) reactions are catalyzed within the living bacterial cell by integrated enzyme systems, the end result being self-replication of the cell. The capability of microbial cells to live, function, and replicate in an appropriate chemical milieu (such as a bacterial culture medium) and the chemical changes that result during this transformation constitute the scope of bacterial metabolism.

The bacterial cell is a highly specialized energy transformer. Chemical energy generated by substrate oxidations is conserved by formation of high-energy compounds such as adenosine diphosphate (ADP) and adenosine triphosphate (ATP) or compounds containing the thioester bond (acetyl ~ SCoA) or succinyl ~ SCoA.

$$(R\text{—}\overset{\overset{\displaystyle O}{\|}}{C} \sim S - R), such\ as\ acetyl \sim S\text{-coenzyme A}$$

ADP and ATP represent adenosine monophosphate (AMP) plus one and two high-energy phosphates (AMP ~ P and AMP ~ P~ P, respectively); the energy is stored in these compounds as high-energy phosphate bonds. In the presence of proper enzyme systems, these compounds can be used as energy sources to synthesize the new complex organic compounds needed by the cell. All living cells must maintain steady-state biochemical reactions for the formation and use of such high-energy compounds.

Kluyver and Donker recognized that bacterial cells, regardless of species, were in many respects similar chemically to all other living cells. For example, these investigators recognized that hydrogen transfer is a common and fundamental feature of all metabolic processes. Bacteria, like mammalian and plant cells, use ATP or the high-energy phosphate bond (~ P) as the primary chemical energy source. Bacteria also require the B-complex vitamins as functional coenzymes for many oxidation-reduction reactions needed for growth and energy transformation. An organism such as Thiobacillus thiooxidans, grown in a medium containing only sulfur and inorganic salts, synthesizes large amounts of thiamine, riboflavine, nicotinic acid, pantothenic acid, pyridoxine, and biotin. Therefore, Kluyver proposed the unity theory of biochemistry (Die Einheit in der Biochemie), which states that all basic enzymatic reactions which support and maintain life processes within cells of organisms, had more similarities than differences. This concept of biochemical unity stimulated many investigators to use bacteria as model systems for studying related eukaryotic, plant and animal biochemical reactions that are essentially "identical" at the molecular level.

From a nutritional, or metabolic, viewpoint, three major physiologic types of bacteria exist: the heterotrophs (or chemoorganotrophs), the autotrophs (or chemolithotrophs), and the photosynthetic bacteria (or phototrophs).

Table: Nutritional Diversity Exhibited by Physiologically Different Bacteria.

Required Components for Bacterial Growth				
Physiology Type	Carbon Source	Nitrogen Source	Energy Source	Hydrogen Source
	Organic	Organic or Inorganic	Oxidation of Inorganic compounds	–
Autotrophic (chemolithotrophic)	CO_2	Inorganic	Oxidation of Inorganic compounds	–
Photosynthetic Photolithotrophic (Bacterial)	CO_2	Inorganic	Sunlight	H_2S or H_2
Cyanobacteria	CO_2	Inorganic	Sunlight	Photolysis of H_2O
Photoorganotrophic (Bacteria)	CO_2	Inorganic	Sunlight	Organic Compounds

Heterotrophic Metabolism

Heterotrophic bacteria, which include all pathogens, obtain energy from oxidation of organic compounds. Carbohydrates (particularly glucose), lipids, and protein are the most commonly oxidized compounds. Biologic oxidation of these organic compounds by bacteria results in synthesis of ATP as the chemical energy source. This process also permits generation of simpler organic compounds (precursor molecules) needed by the bacteria cell for biosynthetic or assimilatory reactions.

The Krebs cycle intermediate compounds serve as precursor molecules (building blocks) for the energy-requiring biosynthesis of complex organic compounds in bacteria. Degradation reactions that simultaneously produce energy and generate precursor molecules for the biosynthesis of new cellular constituents are called amphibolic.

All heterotrophic bacteria require preformed organic compounds. These carbon- and nitrogen-containing compounds are growth substrates, which are used aerobically or anaerobically to generate reducing equivalents (e.g., reduced nicotinamide adenine dinucleotide; $NADH + H^+$); these reducing equivalents in turn are chemical energy sources for all biologic oxidative and fermentative systems. Heterotrophs are the most commonly studied bacteria; they grow readily in media containing carbohydrates, proteins, or other complex nutrients such as blood. Also, growth media may be enriched by the addition of other naturally occurring compounds such as milk (to study lactic acid bacteria) or hydrocarbons (to study hydrocarbon-oxidizing organisms).

Respiration

Glucose is the most common substrate used for studying heterotrophic metabolism. Most aerobic organisms oxidize glucose completely by the following reaction equation:

$$C_6H_{12}O_6 + 6O_2 \rightarrow 6CO_2 + 6H_2O + \text{energy}$$

This equation expresses the cellular oxidation process called respiration. Respiration occurs within the cells of plants and animals, normally generating 38 ATP molecules (as energy) from the oxidation of 1 molecule of glucose. This yields approximately 380,000 calories (cal) per mode of glucose (ATP ~ 10,000 cal/mole). Thermodynamically, the complete oxidation of one mole of glucose should yield approximately 688,000 cal; the energy that is not conserved biologically as chemical energy (or ATP formation) is liberated as heat (308,000 cal). Thus, the cellular respiratory process is at best about 55% efficient.

Glucose oxidation is the most commonly studied dissimilatory reaction leading to energy production or ATP synthesis. The complete oxidation of glucose may involve three fundamental biochemical pathways. The first is the glycolytic or Embden-Meyerhof-Parnas pathway, the second is the Krebs cycle (also called the citric acid cycle or tricarboxylic acid cycle), and the third is the series of membrane-bound electron transport oxidations coupled to oxidative phosphorylation.

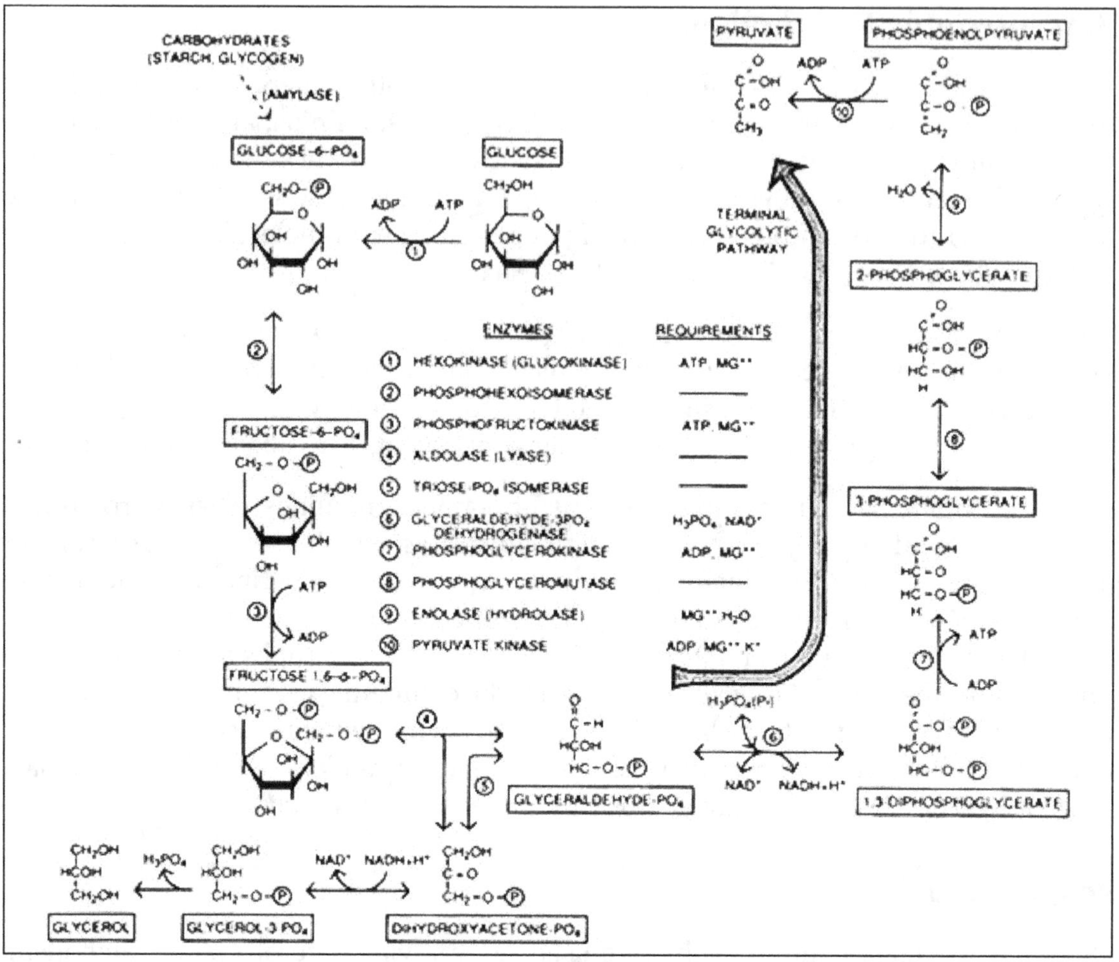

Figure: Glycolytic (EMP) pathway.

Respiration takes place when any organic compound (usually carbohydrate) is oxidized completely to CO_2 and H_2O. In aerobic respiration, molecular O_2 serves as the terminal acceptor of electrons. For anaerobic respiration, NO_3^-, SO_4^{2-}, CO_2, or fumarate can serve as terminal electron acceptors (rather than O_2), depending on the bacterium studied. The end result of the respiratory process is the complete oxidation of the organic substrate molecule, and the end products formed are primarily CO_2 and H_2O. Ammonia is formed also if protein (or amino acid) is the substrate oxidized. The biochemical pathways normally involved in oxidation of various naturally occurring organic compounds.

Metabolically, bacteria are unlike cyanobacteria (blue-green algae) and eukaryotes in that glucose oxidation may occur by more than one pathway. In bacteria, glycolysis represents one of several pathways by which bacteria can catabolically attack glucose. The glycolytic pathway is most commonly associated with anaerobic or fermentative metabolism in bacteria and yeasts. In bacteria, other minor heterofermentative pathways, such as the phosphoketolase pathway, also exist.

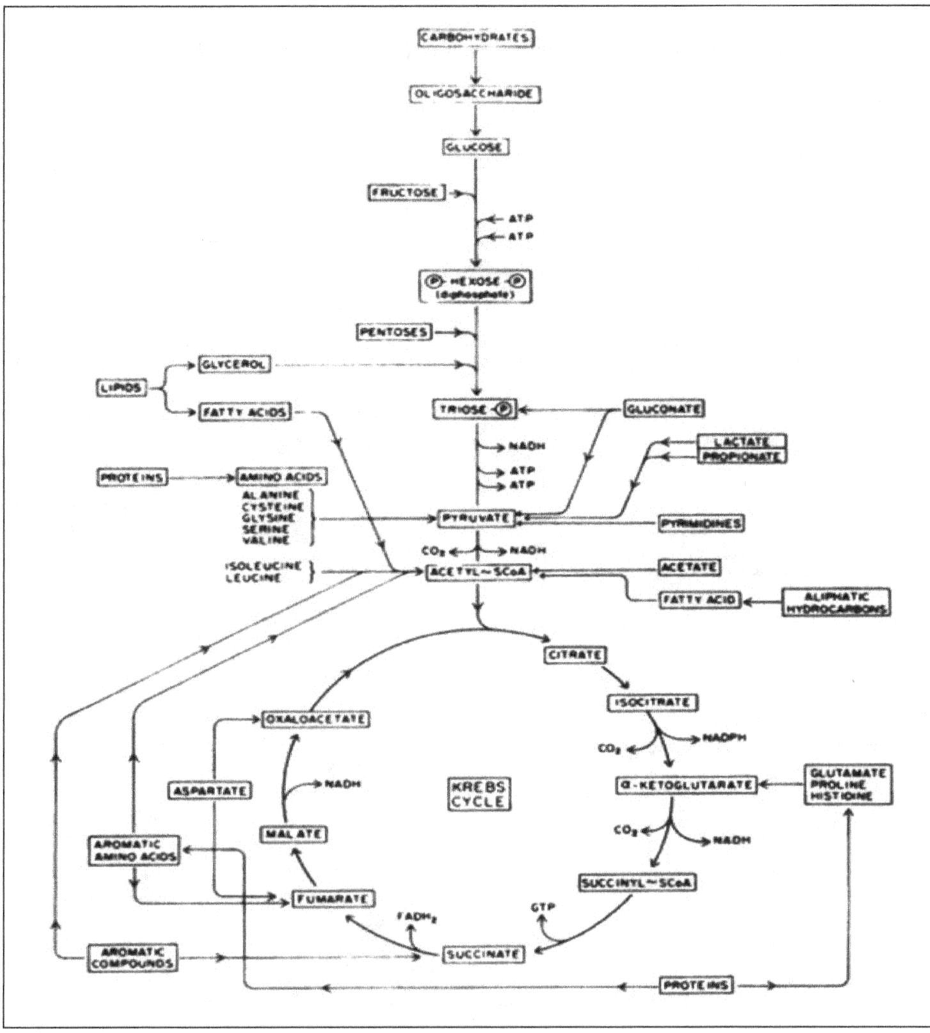

Figure: Heterotrophic metabolism, general pathway.

In addition, two other glucose-catabolizing pathways are found in bacteria: the oxidative pentose phosphate pathway (hexose monophosphate shunt), and the Entner-Doudoroff pathway, which is almost exclusively found in obligate aerobic bacteria. The highly oxidative Azotobacter and most Pseudomonas species, for example, utilize the Entner-Doudoroff pathway for glucose catabolism, because these organisms lack the enzyme phosphofructokinase and hence cannot synthesize fructose 1,6-diphosphate, a key intermediate compound in the glycolytic pathway. (Phospho-fructokinase is also sensitive to molecular O_2 and does not function in obligate aerobes). Other bacteria, which lack aldolase (which splits fructose-1,6-diphosphate into two triose phosphate compounds), also cannot have a functional glycolytic pathway. Although the Entner-Doudoroff pathway is usually associated with obligate aerobic bacteria, it is present in the facultative anaerobe Zymomonas mobilis (formerly Pseudomonas lindneri). This organism dissimilates glucose to ethanol and represents a major alcoholic fermentation reaction in a bacterium.

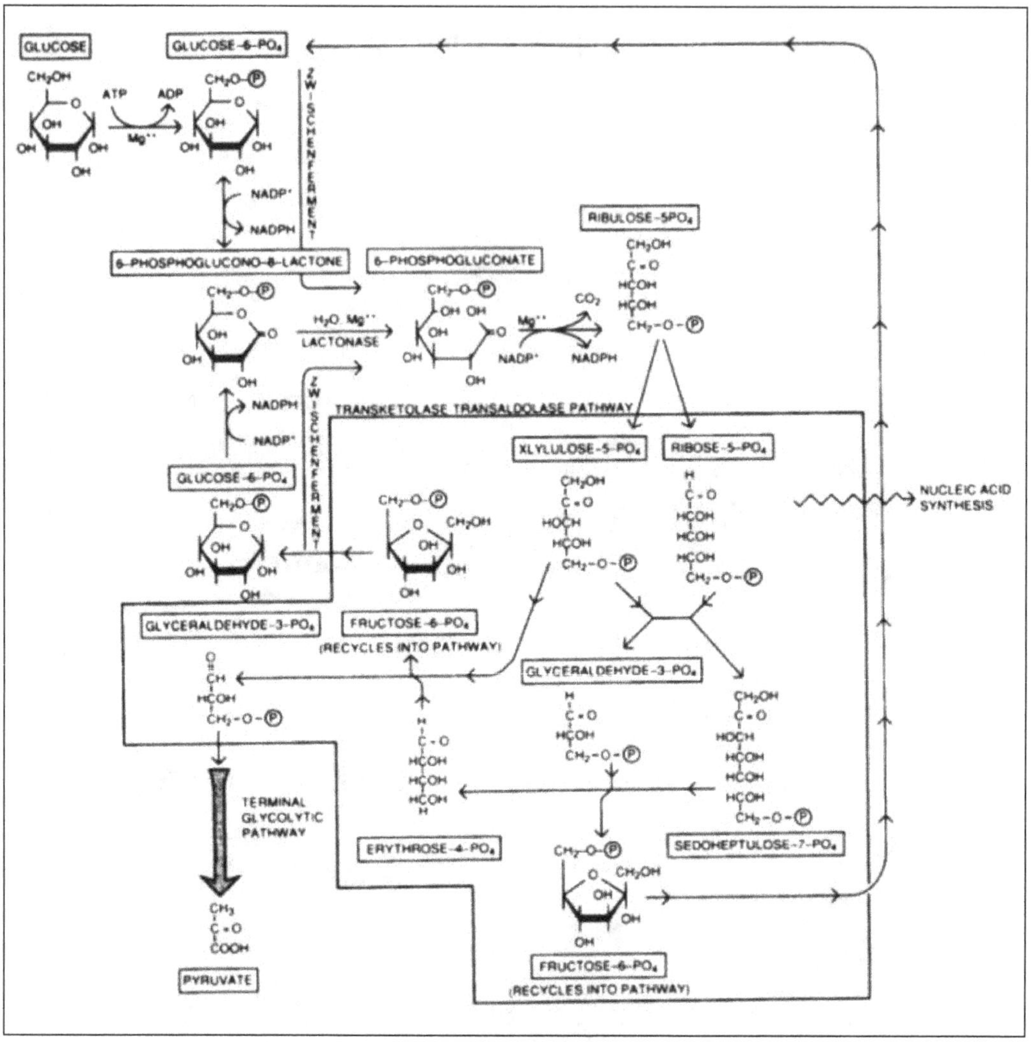

Figure: Hexose monophosphate (HMS) pathway.

Glucose dissimilation also occurs by the hexose monophosphate shunt. This oxidative pathway was discovered in tissues that actively metabolize glucose in the presence of two glycolytic pathway inhibitors (iodoacetate and fluoride). Neither inhibitor had an effect on glucose dissimilation, nor did NADPH + H^+ generation occur directly from the oxidation of glucose-6-phosphate (to 6-phosphoglucono-δ-lactone) by glucose-6phosphate dehydrogenase. The pentose phosphate pathway subsequently permits the direct oxidative decarboxylation of glucose to pentoses. The capability of this oxidative metabolic system to bypass glycolysis explains the term shunt.

The biochemical reactions of the Entner-Doudoroff pathway are a modification of the hexose monophosphate shunt, except that pentose sugars are not directly formed. The two pathways are identical up to the formation of 6-phosphogluconate and then diverge. In the Entner-Doudoroff pathway, no oxidative decarboxylation of 6-phosphogluconate occurs and no pentose compound is formed. For this pathway, a new 6 carbon compound intermediate (2-keto-3-deoxy6-phosphogluconate) is generated by the action of 6-phosphogluconate dehydratase (an Fe^{2+}– and glutathione-stimulated enzyme); this intermediate compound is then directly cleaved into the triose (pyruvate) and a triose-phosphate compound (glyceraldehyde-3-phosphate) by the 2-keto-3-deoxy6-phosphogluconate aldolase. The glyceraldehyde-3-phosphate is further oxidized to another pyruvate molecule by the same enzyme systems that catalyze the terminal glycolytic pathway.

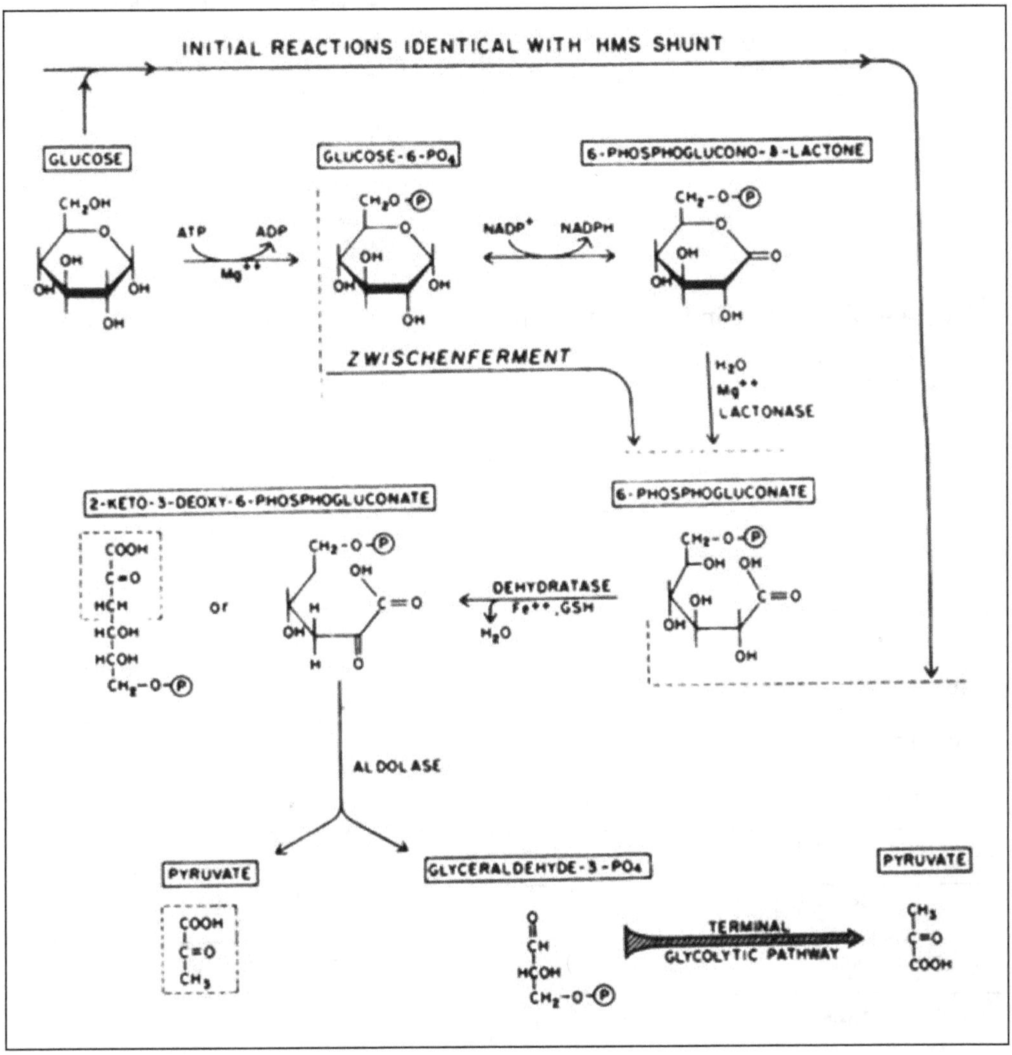

Figure: Entner-Doudoroff (ED) pathway.

The glycolytic pathway may be the major one existing concomitantly with the minor oxidative pentose phosphate - hexose monophosphate shunt pathway; the Entner-Doudoroff pathway also may function as a major pathway with a minor hexose monophosphate shunt. A few bacteria possess only one pathway. All cyanobacteria, Acetobacter suboxydans, and A. xylinum possess only the hexose monophosphate shunt pathway; Pseudomonas saccharophilia and Z. mobilis possess solely the

Entner-Doudoroff pathway. Thus, the end products of glucose dissimilatory pathways are as follows:

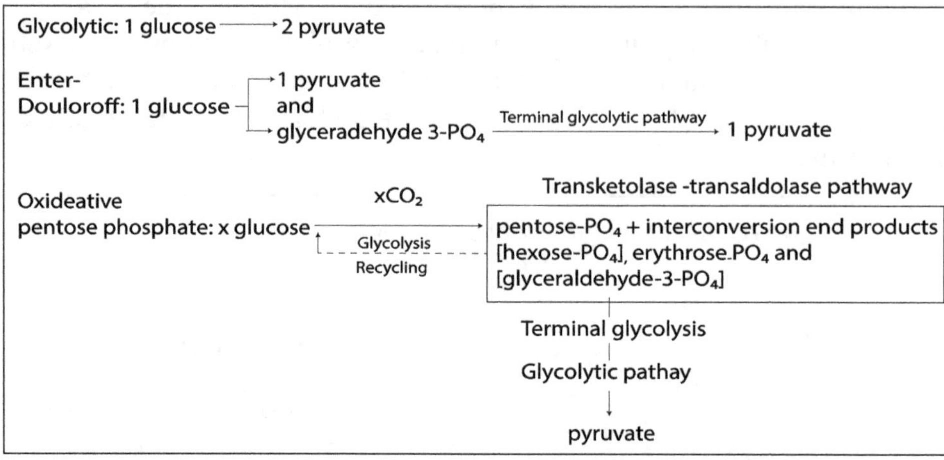

Table: Glucose Dissimilation Pathways Utilized by Bacteria, Cyanobacteria, and Yeasts.

Bacteria	Glycolytic Pathway	Oxidative Pentose Phosphate Pathway	Entner- Doudoroff Pathway
Acetobacter suboxydans		Sole	
Acetobacter Xylinum		Sole	
Acetobacterium			Major
Agrobacterium spp			Major
Bacillus subtilis	Major	Minor	
Caulobacter spp			Major
Escherichia coli	Major		Minor
Lactobacillus delbrueckii	Major		
leuconostoc mesenteroides		Minor	
Neisseria gonorrhoeae		Minor	Major
Neisseria meningitidis		Minor	Major
Neisseria perflava		Minor	
Neisseria sicca		Major	
Pseudomonas aeruginosa			Major
Pseudomonas saccharophila			Sole
Rhizobium spp			Major

Sarcina lutea	Major	Minor	
Spirillum spp			Major
Streptococcus faecalis			(Major)
Streptomyces griseus	Major	Minor	
zymomonas anaerobia			Sole
Zymomonas mobilis			Sole
All Cyanobacteria			Sole
All yeast	Major	Minor	

All major pathways of glucose or hexose catabolism have several metabolic features in common. First, there are the preparatory steps by which key intermediate compounds such as the triose-PO4, glyceraldehyde-3-phosphate, and/or pyruvate are generated. The latter two compounds are almost universally required for further assimilatory or dissimilatory reactions within the cell. Second, the major source of phosphate for all reactions involving phosphorylation of glucose or other hexoses is ATP, not inorganic phosphate (Pi). Actually, chemical energy contained in ATP must be initially spent in the first step of glucose metabolism (via kinase-type enzymes) to generate glucose-6-phosphate, which initiates the reactions involving hexose catabolism. Third, NADH + H^+ or NADPH + H^+ is generated as reducing equivalents (potential energy) directly by one or more of the enzymatic reactions involved in each of these pathways.

Fermentation

Fermentation, another example of heterotrophic metabolism, requires an organic compound as a terminal electron (or hydrogen) acceptor. In fermentations, simple organic end products are formed from the anaerobic dissimilation of glucose (or some other compound). Energy (ATP) is generated through the dehydrogenation reactions that occur as glucose is broken down enzymatically. The simple organic end products formed from this incomplete biologic oxidation process also serve as final electron and hydrogen acceptors. On reduction, these organic end products are secreted into the medium as waste metabolites (usually alcohol or acid). The organic substrate compounds are incompletely oxidized by bacteria, yet yield sufficient energy for microbial growth. Glucose is the most common hexose used to study fermentation reactions.

For most microbial fermentations, glucose dissimilation occurs through the glycolytic pathway. The simple organic compound most commonly generated is pyruvate, or a compound derived enzymatically from pyruvate, such as acetaldehyde, α-acetolactate, acetyl ~ SCoA, or lactyl ~ SCoA. Acetaldehyde can then be reduced by NADH + H^+ to ethanol, which is excreted by the cell. The end product of lactic acid fermentation, which occurs in streptococci (e.g., Streptococcus lactis) and many lactobacilli (e.g., Lactobacillus casei, L. pentosus), is a single organic acid, lactic acid. Organisms that produce only lactic acid from glucose fermentation are homofermenters. Homofermentative lactic acid bacteria dissimilate glucose exclusively through the glycolytic pathway. Organisms that ferment glucose to multiple end products, such as acetic acid, ethanol, formic acid, and CO_2, are referred to as heterofermenters. Examples of heterofermentative bacteria include Lactobacillus, Leuconostoc, and Microbacterium species. Heterofermentative fermentations are more common among bacteria, as in the mixed-acid fermentations carried out by bacteria of the family Enterobacteriaceae (e.g., Escherichia coli, Salmonella, Shigella, and Proteus species).

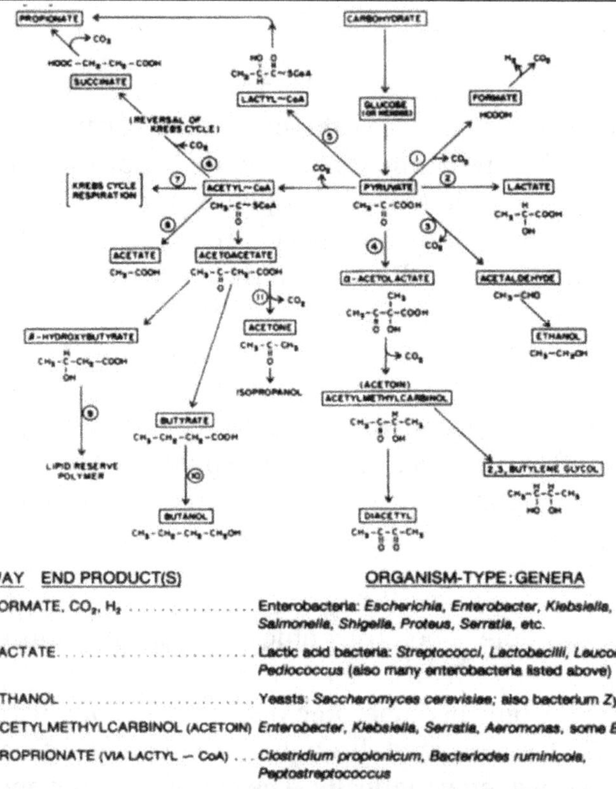

PATHWAY	END PRODUCT(S)	ORGANISM-TYPE: GENERA
1	FORMATE, CO_2, H_2	Enterobacteria: *Escherichia, Enterobacter, Klebsiella, Salmonella, Shigella, Proteus, Serratia, etc.*
2	LACTATE	Lactic acid bacteria: *Streptococci, Lactobacilli, Leuconostoc, Pediococcus* (also many enterobacteria listed above)
3	ETHANOL	Yeasts: *Saccharomyces cerevisiae;* also bacterium *Zymomona*
4	ACETYLMETHYLCARBINOL (ACETOIN)	*Enterobacter, Klebsiella, Serratia, Aeromonas,* some *Bacillus*
5	PROPRIONATE (VIA LACTYL – CoA)	*Clostridium propionicum, Bacteriodes ruminicola, Peptostreptococcus*
6	PROPIONATE (VIA SUCCINATE)	Propionic acid bacteria: *Propionibacterium, Vielonella*
7		obligately aerobic bacteria
8		*Acetobacter, Gluconobacter*
9		most bacterial species
10	BUTANOL (VIA BUTYRATE)	*Clostridium acetobutylicum, Clostridium butyricum*
11	ISOPROPANOL (VIA ACETONE)	*Clostridium butyricum*

Figure: Fermentative pathways of bacteria and the major end products formed with the organism type carrying out the fermentation.

Many of these glucose fermenters usually produce CO_2 and H_2 with different combinations of acid end products (formate, acetate, lactate, and succinate). Other bacteria such as Enterobacter aerogenes, Aeromonas, Serratia, Erwinia, and Bacillus species also form CO_2 and H_2 as well as other neutral end products (ethanol, acetylmethylcarbinol [acetoin], and 2,3-butylene glycol). Many obligately anaerobic clostridia (e.g.,

Clostridium saccharobutyricum, C. thermosaccharolyticum) and Butyribacterium species ferment glucose with the production of butyrate, acetate, CO_2, and H_2, whereas other Clostridum species (C. acetobutylicum and C. butyricum) also form these fermentation end products plus others (butanol, acetone, isopropanol, formate, and ethanol). Similarly, the anaerobic propionic acid bacteria (Propionibacterium species) and the related Veillonella species ferment glucose to form CO_2, propionate, acetate, and succinate. In these bacteria, propionate is formed by the partial reversal of the Krebs cycle reactions and involves a CO_2 fixation by pyruvate (the Wood-Werkman reaction) that forms oxaloacetate (a four-carbon intermediate). Oxaloacetate is then reduced to malate, fumarate, and succinate, which is decarboxylated to propionate. Propionate is also formed by another three-carbon pathway in C. propionicum, Bacteroides ruminicola, and Peptostreptococcus species, involving a lactyl ~ SCoA intermediate. The obligately aerobic acetic acid bacteria (Acetobacter and the related Gluconobacter species) can also ferment glucose, producing acetate and gluconate. Below figure summarizes the pathways by which the various major fermentation end products form from the dissimilation of glucose through the common intermediate pyruvate.

Table: Energy Obtained from Bacterial Fermentations by Substrate Phosphorylations.

	Fermentation	Actual Energy (cal/mole)	Theoretical Energy (cal/mole)	Efficiency (%)
Homolactic				
$C_6H_{12}O_6$ (Glucose)	$\xrightarrow{\text{Glycolysis}}$ $2CH_3-C(H)(OH)-COOH$ + (Lactic acid)	-20,000	57,000	35
Alconolic				
$C_6H_{12}O_6$ (Glucose)	$\xrightarrow{\text{Glycolysis}}$ $2CH_3-C(H)(H)-CH + 2CO_2$ + (Ethanol)	-20,000	58,000	34

For thermodynamic reasons, bacteria that rely on fermentative process for growth cannot generate as much energy as respiring cells. In respiration, 38 ATP molecules (or approximately 380,000 cal/mole) can be generated as biologically useful energy from the complete oxidation of 1 molecule of glucose (assuming 1 NAD(P)H = 3 ATP and 1 ATP → ADP + Pi = 10,000 cal/mole). Table shows comparable bioenergetic parameters for the lactate and ethanolic fermentations by the glycolytic pathway. Although only 2 ATP molecules are generated by this glycolytic pathway, this is apparently enough energy to permit anaerobic growth of lactic acid bacteria and the ethanolic fermenting yeast, Saccharomyces cerevisiae. The ATP-synthesizing reactions in the glycolytic pathway specifically involve the substrate phosphorylation reactions catalyzed by phosphoglycerokinase and pyruvic kinase. Although all the ATP molecules available for fermentative growth are believed to be generated by these substrate phosphorylation reactions, some energy equivalents are also generated by proton extrusion reactions (acid liberation), which occur with intact membrane systems and involve the proton extrusion reactions of energy conservation as it applies to fermentative metabolism.

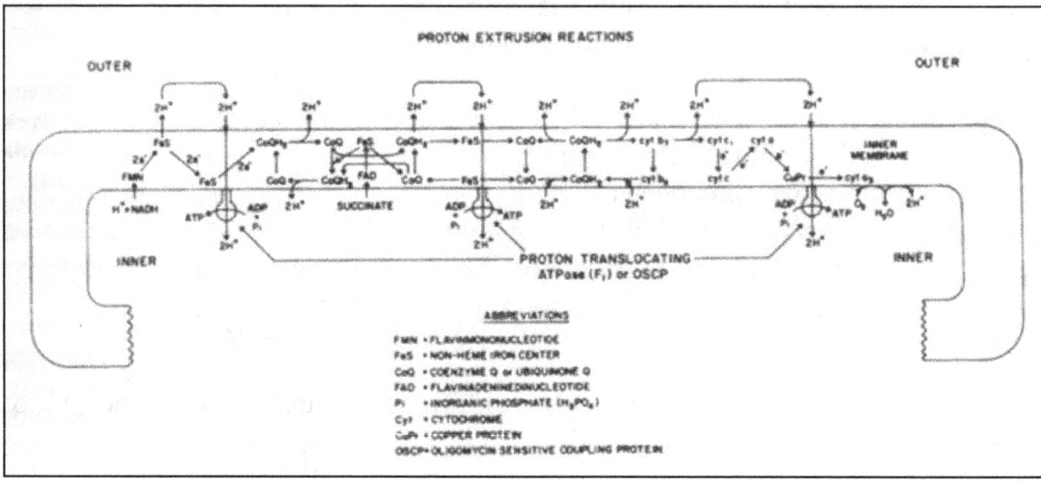

Figure: Mitchell hypotheses, a chemiosmotic model of energy transduction.

Krebs Cycle

The Krebs cycle (also called the tricarboxylic acid cycle or citic acid cycle) functions oxidatively in respiration and is the metabolic process by which pyruvate or acetyl ~ SCoA is completely decarboxylated to CO_2. In bacteria, this reaction occurs through acetyl ~ SCoA, which is the first product in the oxidative decarboxylation of pyruvate by pyruvate dehydrogenase. Bioenergetically, the following overall exergonic reaction occurs:

$$CH_3-\underset{\underset{O}{\|}}{C}-COOH+5O \xrightarrow[\textit{Electron transport and oxidative phosphorviation}]{\textit{Krebs cycle (vi2 } CH_3-\overset{\overset{O}{\|}}{C}-O-SCoA}$$

$$3Co_2+2H_2O+15ATP(\approx 150{,}000\,cal/mole)$$

If 2 pyruvate molecules are obtained from the dissimilation of 1 glucose molecule, then 30 ATP molecules are generated in total. The decarboxylation of pyruvate, isocitrate, and α-ketoglutarate accounts for all CO_2 molecules generated during the respiratory process. Figure shows the enzymatic reactions in the Krebs cycle. The chemical energy conserved by the Krebs cycle is contained in the reduced compounds generated (NADH + H^+, NADPH + H^+, and succinate). The potential energy inherent in these reduced compounds is not available as ATP until the final step of respiration (electron transport and oxidative phosphorylation) occurs.

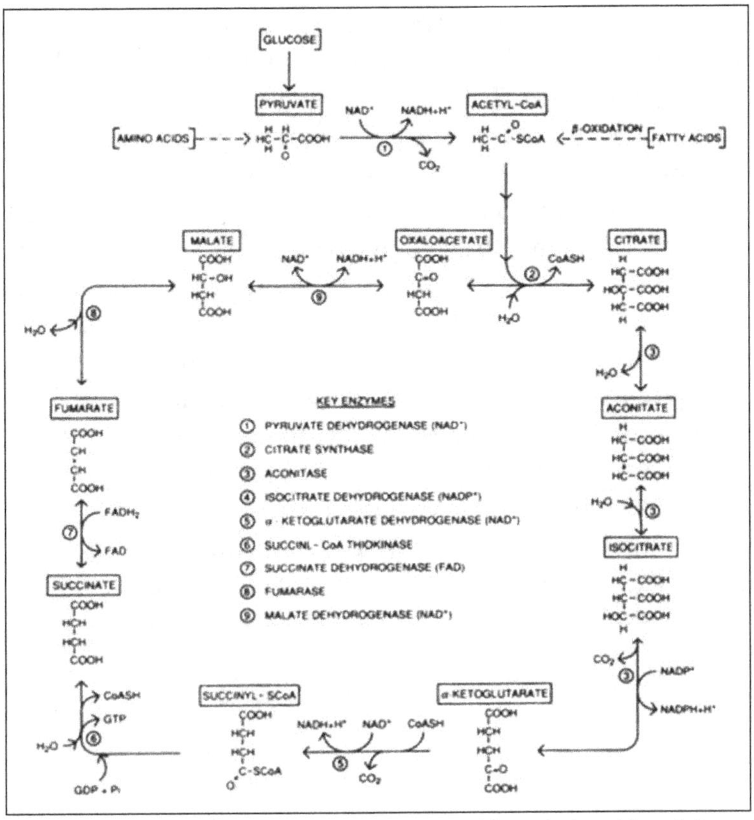

Figure: Krebs cycle (also tricarboxylic acid or citric acid cycle).

The Krebs cycle is therefore another preparatory stage in the respiratory process. If 1 molecule of pyruvate is oxidized completely to 3 molecules of CO_2, generating 15 ATP molecules, the oxidation of 1 molecule of glucose will yield as many as 38 ATP molecules, provided glucose is dissimilated by glycolysis and the Krebs cycle (further assuming that the electron transport/oxidative phosphorylation reactions are bioenergetically identical to those of eukaryotic mitochondria).

Glyoxylate Cycle

In general, the Krebs cycle functions similarly in bacteria and eukaryotic systems, but major differences are found among bacteria. One difference is that in obligate aerobes, L-malate may be oxidized directly by molecular O_2 via an electron transport chain. In other bacteria, only some Krebs cycle intermediate reactions occur because α-ketoglutarate dehydrogenase is missing.

A modification of the Krebs cycle, commonly called the glyoxylate cycle, or shunt, which exists in some bacteria. This shunt functions similarly to the Krebs cycle but lacks many of the Krebs cycle enzyme reactions. The glyoxylate cycle is primarily an oxidative pathway in which acetyl~SCoA is generated from the oxidation, of acetate, which usually is derived from the oxidation of fatty acids. The oxidation of fatty acids to acetyl~SCoA is carried out by the β-oxidation pathway. Pyruvate oxidation is not directly involved in the glyoxylate shunt, yet this shunt yields sufficient succinate and malate, which are required for energy production. The glyoxylate cycle also generates other precursor compounds needed for biosynthesis. The glyoxylate cycle was discovered as an unusual metabolic pathway during an attempt to learn how lipid (or acetate) oxidation in bacteria and plant seeds could lead to the direct biosynthesis of carbohydrates. The glyoxylate cycle converts oxaloacetate either to pyruvate and CO_2 (catalyzed by pyruvate carboxylase) or to phosphoenolpyruvate and CO_2 (catalyzed by the inosine triphosphate [ITP]-dependent phosphoenolpyruvate carboxylase kinase). Either triose compound can then be converted to glucose by reversal of the glycolytic pathway. The glyoxylate cycle is found in many bacteria, including Azotobacter vinelandii and particularly in organisms that grow well in media in which acetate and other Krebs cycle dicarboxylic acid intermediates are the sole carbon growth source. One primary function of the glyoxylate cycle is to replenish the tricarboxylic and dicarboxylic acid intermediates that are normally provided by the Krebs cycle. A pathway whose primary purpose is to replenish such intermediate compounds is called anaplerotic.

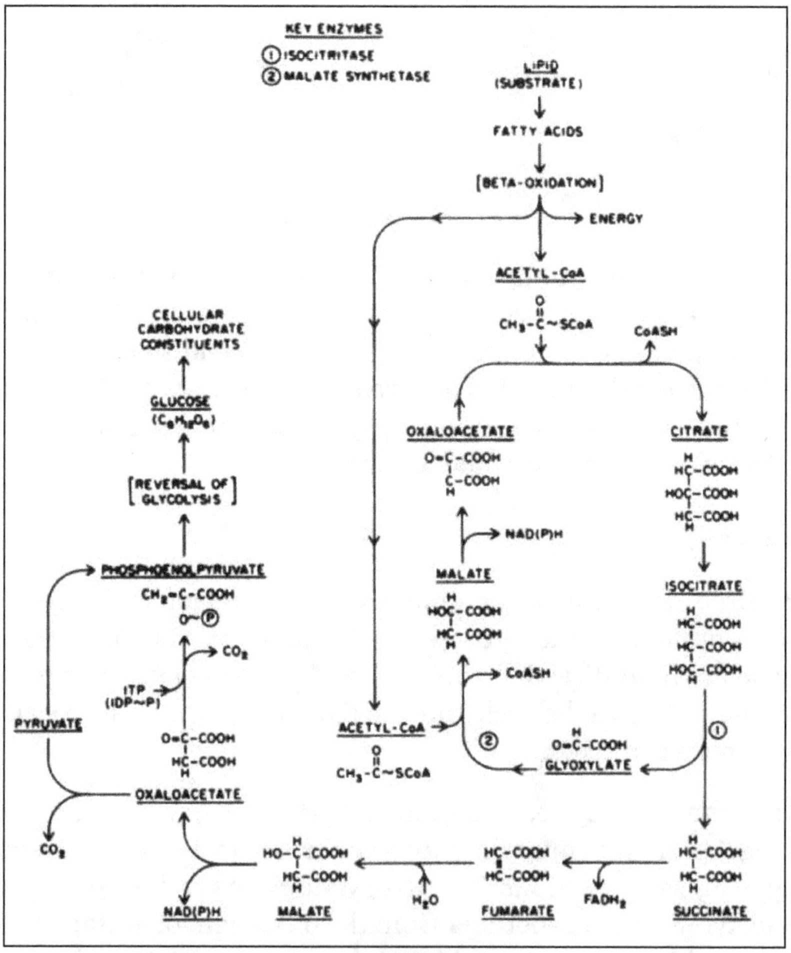

Figure: Glyoxylate shunt.

Electron Transport and Oxidative Phosphorylation

The final stage of respiration occurs through a series of oxidation-reduction electron transfer reactions that yield the energy to drive oxidative phosphorylation; this in turn produces ATP. The enzymes involved in electron transport and oxidative phosphorylation reside on the bacterial inner (cytoplasmic) membrane. This membrane is invaginated to form structures called respiratory vesicles, lamellar vesicles, or mesosomes, which function as the bacterial equivalent of the eukaryotic mitochondrial membrane.

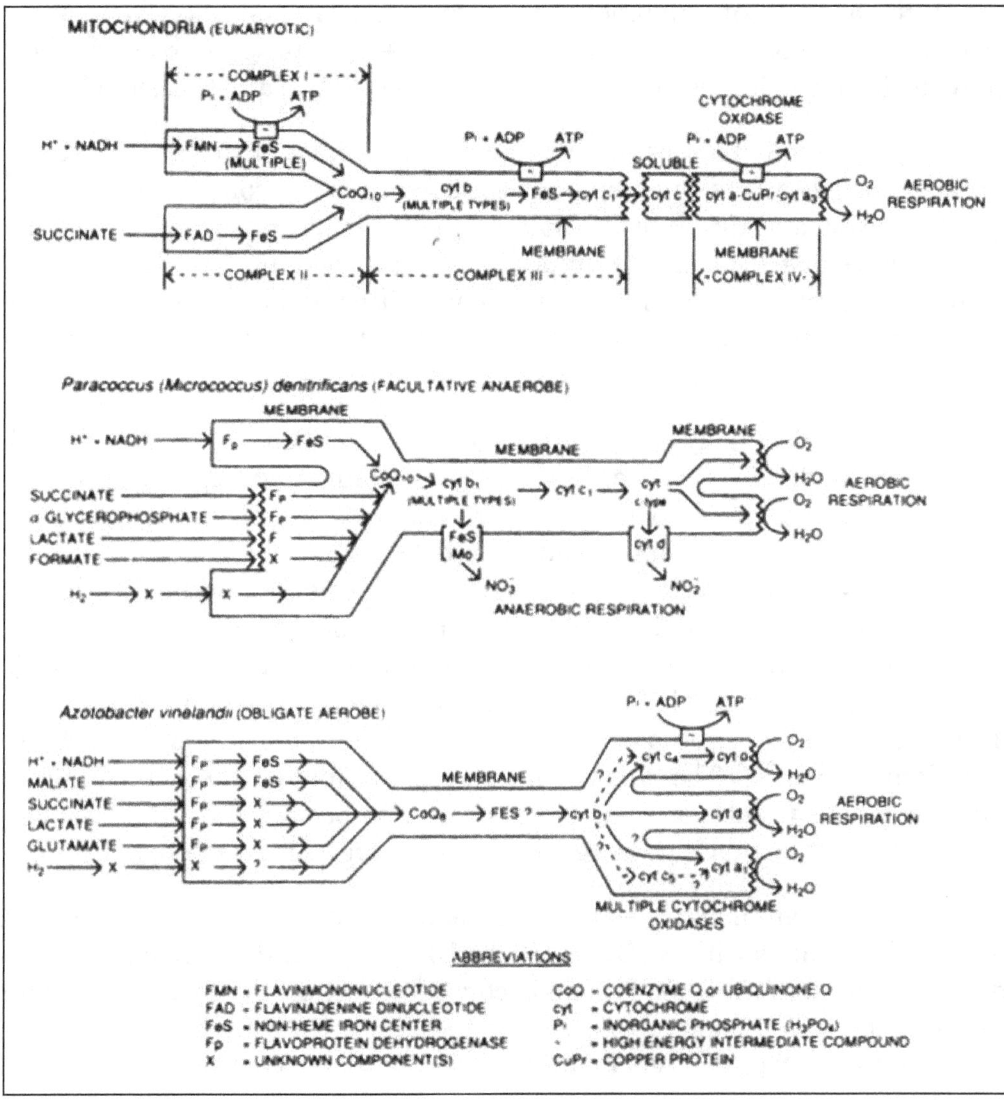

Figure: Respiratory electron transport chains.

Respiratory electron transport chains vary greatly among bacteria, and in some organisms are absent. The respiratory electron transport chain of eukaryotic mitochondria oxidizes NADH + H^+, NADPH + H^+, and succinate (as well as the coacylated fatty acids such as acetyl~SCoA). The bacterial electron transport chain also oxidizes these compounds, but it can also directly oxidize, via non-pyridine nucleotide-dependent pathways, a larger variety of reduced substrates such as lactate, malate, formate, α-glycerophosphate, H^2, and glutamate. The respiratory electron carriers in bacterial electron transport systems are more varied than in eukaryotes, and the chain is usually branched at the site(s) reacting with molecular O_2. Some electron carriers, such as nonheme iron centers and ubiquinone (coenzyme Q), are common to both the bacterial and mammalian respiratory electron transport chains. In some bacteria, the naphthoquinones or vitamin K may be found with ubiquinone. In still other bacteria, vitamin K serves in the absence of ubiquinone. In mitochondrial respiration, only one cytochrome oxidase component is found (cytochrome a + a_3 oxidase). In bacteria there are multiple cytochrome oxidases, including cytochromes a, d, o, and occasionally a + a_3.

In bacteria cytochrome oxidases usually occur as combinations of $a_{1:\,d:\,o}$ and a + $a_{3:o}$. Bacteria also possess mixed-function oxidases such as cytochromes P-450 and P-420 and cytochromes c' and c'c', Which also react with carbon monoxide. These diverse types of oxygen-reative cytochromes undoubtedly have evolutionary significance. Bacteria were present before O_2 was formed; when O_2 became available as a metabolite, bacteria evolved to use it in different ways; this probably accounts for the diversity in bacterial oxygen-reactive hemoproteins.

Cytochrome oxidases in many pathogenic bacteria are studied by the bacterial oxidase reaction, which subdivides Gram-negative organisms into two major groups, oxidase positive and oxidase negative. This oxidase reaction is assayed for by using N,N,N', N'-tetramethyl-p-phenylenediamine oxidation (to Wurster's blue) or by using indophenol blue synthesis (with dimethyl-p-phenylenediamine and α-naphthol). Oxidase-positive bacteria contain integrated (cytochrome c type: oxidase) complexes, the oxidase component most frequently encountered is cytochrome o, and occasionally a + a_3. The cytochrome oxidase responsible for the indophenol oxidase reaction complex was isolated from membranes of Azotobacter vinelandii, a bacterium with the highest respiratory rate of any known cell. The cytochrome oxidase was found to be an integrated cytochrome $c_{4:o}$ complex, which was shown to be present in Bacillus species. These Bacillus strains are also highly oxidase positive, and most are found in morphologic group II.

Both bacterial and mammalian electron transfer systems can carry out electron transfer (oxidation) reactions with NADH + H^+, NADPH + H^+, and succinate. Energy generated from such membrane oxidations is conserved within the membrane and then transferred in a coupled manner to drive the formation of ATP. The electron transfer sequence is accomplished entirely by membrane-bound enzyme systems. As the electrons are transferred by a specific sequence of electron carriers, ATP is synthesized from ADP + inorganic phosphate (Pi) or orthophosphoric acid (H_3PO_4).

In respiration, the electron transfer reaction is the primary mode of generating energy; electrons (2e-) from a low-redox-potential compound such as NADH + H^+ are sequentially transferred to a specific flavoprotein dehydrogenase or oxidoreductase (flavin mononucleotide [FMN] type for NADH or flavin adenine dinucleotide [FAD] type for succinate); this electron pair is then transferred to a nonheme iron center (FeS) and finally to a specific ubiquinone or a naphthoquinone derivative. This transfer of electrons causes a differential chemical redox potential change so that within the membrane enough chemical energy is conserved to be transferred by a coupling mechanism to a high-energy compound (e.g., ADP + Pi $\rightarrow$ ATP). ATP molecules represent the final stable high-energy intermediate compound formed.

A similar series of redox changes also occurs between ubiquinone and cytochrome c, but with a greater differential in the oxidation-reduction potential level, which allows for another ATP synthesis step. The final electron transfer reaction occurs at the cytochrome oxidase level between reduced cyotchrome c and molecular O_2; this reaction is the terminal ATP synthesis step.

Mitchell or Proton Extrusion Hypothesis

A highly complex but attractive theory to explain energy conservation in biologic systems is the chemiosmotic coupling of oxidative and photosynthetic phosphorylations, commonly called the Mitchell hypothesis. This theory attempts to explain the conservation of free energy in this process on the basis of an osmotic potential caused by a proton concentration differential (or proton gradient) across a proton-impermeable membrane. Energy is generated by a proton extrusion reaction during membrane-bound electron transport, which in essence serves as a proton pump; energy conservation and coupling follow. This represents an obligatory "intact" membrane phenomenon. The energy thus conserved (again within the confines of the membrane and is coupled to ATP synthesis. This would occur in all biologic cells, even in the lactic acid bacteria that lack a cytochrome-dependent electron transport chain but still possesses a cytoplasmic membrane. In this hypothesis, the membrane allows for charge separation, thus forming a proton gradient that drives all bioenergization reactions. By such means, electromotive forces can be generated by oxidation-reduction reactions that can be directly coupled to ion translocations, as in the separation of H^+ and OH^- ions in electrochemical systems. Thus, an enzyme or an electron transfer carrier on a membrane that undergoes an oxidation-reduction reaction serves as a specific conductor for OH^- (or o^{2-}), and "hydrodehydration" provides electromotive power, as it does in electrochemical cells.

The concept underlying Mitchell's hypothesis is complex, and many modifications have been proposed, but the theory's most attractive feature is that it unifies all bioenergetic conservation principles into a single concept requiring an intact membrane vesicle to function properly. Shows how the Mitchell hypothesis might be used to explain energy generation, conservation, and transfer by a coupling process. The least satisfying aspect of the chemiosmotic hypothesis is the lack of understanding of how chemical energy is actually conserved within the membrane and how it is transmitted by coupling for ATP synthesis.

Bacterial Photosynthesis

Many prokaryotes (bacteria and cyanobacteria) possess phototrophic modes of metabolism. The types of photosynthesis in the two groups of prokaryotes differ mainly in the type of compound that serves as the hydrogen donor in the reduction of CO_2 to glucose. Phototrophic organisms differ from heterotrophic organisms in that they utilize the glucose synthesized intracellularly for biosynthetic purposes (as in starch synthesis) or for energy production, which usually occurs through cellular respiration.

Unlike phototrophs, heterotrophs require glucose (or some other preformed organic compound) that is directly supplied as a substrate from an exogenous source. Heterotrophs cannot synthesize large concentrations of glucose from CO_2 by specifically using H_2O or (H_2S) as a hydrogen source and sunlight as energy. Plant metabolism is a classic example of photolithotrophic metabolism: plants need CO_2 and sunlight; H_2O must be provided as a hydrogen source and usually NO_3^- is the nitrogen source for protein synthesis. Organic nitrogen, supplied as fertilizer, is converted to NO_3^- in all soils by bacteria via the process of ammonification and nitrification. Although plant cells are phototrophic, they also exhibit a heterotrophic mode of metabolism in that they respire. For example, plants use classic respiration to catabolize glucose that is generated photosynthetically. Mitochondria as well as the soluble enzymes of the glycolytic pathway are required for glucose dissimilation, and these enzymes are also found in all plant cells. The soluble Calvin cycle enzymes, which are required for glucose synthesis during photosynthesis, are also found in plant cells. It is not possible to feed a plant by pouring a glucose solution on it, but water supplied to a plant will be "photolysed" by chloroplasts in the presence of light; the hydrogen(s) generated from H_2O is used by Photosystems I and II (PSI and PSII) to reduce $NADP^+$ to $NADPH+ H^+$. With the ATP generated by PSI and PSII, these reduced pyridine nucleotides, CO_2 is reduced intracellularly to glucose. This metabolic process is carried out in an integrated manner by Photosystems I and II ("Z" scheme) and by the Calvin cycle pathway.

A new photosynthetic, and nitrogen fixing bacterium, Heliobacterium chlorum, staining Gram positive was isolated, characterized, and found to contain a new type of chlorophyll, i.e., bacteriochlorophyll 'g'. 16S r-RNA sequence analyses showed this organism to be phylogenetically related to members of the family Bacillaceae, although all currently known phototrophes are Gram negative. A few Heliobacteriium strains did show the presence of endospores. Another unusual phototrophe is the Gram negative Halobacterium halobium (now named Halobacterium salinarium), an archaebacterium growing best at 30 °C in 4.0–5.0 M (or 25%, w/v) NaCl. This bacterium is a facultative phototrophe having a respiratory mode; it also possesses a purple membrane within which bacteriorhodopsin serves as the active photosynthetic pigment. This purple membranae possesses a light driven proton translocation pump which mediates photosynthetic ATP synthesis via a proton extrusion reaction. Table summarizes the characteristics of known photosynthetic bacteria.

Table: Characteristics Commonly Exhibited by Phototrophic Bacteria.

Photosynthetic Type	Characteristics	Representative Families and Genera
Purple Bacteria. Sulfur-type (formerly thiorho daceae), photoorganotrophic bacteria.	Phototrophs. Strict anaerobes. H_2S (or H_2)serve as H source. Possess S granules when H_2S used. Contain bacteriochlorophyll a or b.	Chromatiaceae (Chromatium, Thiospirillum, Thiosarcina, Thiocapsa).
Non-sulfur-type (formerly Athiorhodaceae), photoorganotrophic bacteria.	Faculative phototrophs (have respiratory mechanism and will grow heterotrophically). Oxygen-tolerant anaerobes. Most require one or more B vitamins Simple organic compounds serve as H source. Contain bacteriochlorophyll a or b.	Rhodospirillaceae (Rhodopseeudomonas, Rhodomicrobium).
Green bacteria. Photolithotrophic bacteria.	Obligate phototrophs. Strict anaerobes. Contains chlorobium chlorophyll, Which is currently referred to as bacteriochlorophyll type c and d. Many require vitamin B_{12}. S_2 deposited ectracellularly.	Chlorobiaceae (Chlorobium, Chloropseudomonas.

Autotrophy

Bacteria that grow solely at the expense of inorganic compounds (mineral ions), without using sunlight as an energy source, are called autotrophs, chemotrophs, chemoautotrophs, or chemolithotrophs. Like photosynthetic organisms, all autotrophs use CO_2 as a carbon source for growth; their nitrogen comes from inorganic compounds such as NH_3, NO_3^-, or N_2. Interestingly, the energy source for such organisms is the oxidation of specific inorganic compounds. Which inorganic compound is oxidized depends on the bacteria in question. Many autotrophs will not grow on media that contain organic matter, even agar.

Table: Inorganic Oxidation Reactions used by Autotrophic Bacteria as Energy Sources.

Chemosynthetic Type	Inorganic Compounds Oxidized as Energy(-E) Source	Representative Families, Genera, and Species	Nitrogen Cycle Reaction
NH_3 oxidizers (aerobic)	$NH_3 \rightarrow NO_2$ (↓ (–E)) (~E) = Chemocal energy or atp produced	Nitrobacteriaceae (Nitrosomonas, Niotrosococcus, Nitrpspspira)	Nitrification Nitrification Nitrification Nitrification

NO_2 oxidizers (aerobic)	$NO_2 \rightarrow NO_4$ ↓ (–E)	Nitrobacteriaceae (Nitrobacter, Nitrococcus)	Nitrification Nitrification Nitrification
Sulfur oxidizers (aerobic) Iron oxidizers (aerobic)	$S_1 \rightarrow SO_4$ ↓ (–E) $Fe^{2+} \rightarrow Fe^{3+}$ ↓ (–E)	Thiobacillus Thiooxidans, Thiobacillus, terrooxidans, Ferrobacillus, Leptothrrix	
Sulfur-compound oxidizers Denitrification (anaerobic)	S_2o_3 oxidized; NO_3 reduced	Thiobacillus denitrificans	

Also found among the autotrophic microorganisms are the sulfur-oxidizing or sulfur-compound-oxidizing bacteria, which seldom exhibit a strictly autotrophic mode of metabolism like the obligate nitrifying bacteria. The representative sulfur compounds oxidized by such bacteria are H_2S, S_2, and S_2O_3. Among the sulfur bacteria are two very interesting organisms; Thiobacillus ferrooxidans, which gets its energy for autotrophic growth by oxidizing elemental sulfur or ferrous iron, and T. denitrificans, which gets its energy by oxidizing S_2O_3 anaerobically, using NO_3^- as the sole terminal electron acceptor. T denitrificans reduces NO_3 to molecular N_2, which is liberated as a gas; this biologic process is called denitrification. All autotrophic bacteria must assimilate CO_2, which is reduced to glucose from which organic cellular matter is synthesized.

Anaerobic Respiration

Some bacteria exhibit a unique mode of respiration called anaerobic respiration. These heterotrophic bacteria that will not grow anaerobically unless a specific chemical component, which serves as a terminal electron acceptor, is added to the medium. Among these electron acceptors are NO_3^-, SO_4^{2-}, the organic compound fumarate, and CO_2. Bacteria requiring one of these compounds for anaerobic growth are said to be anaerobic respirers.

A large group of anaerobic respirers are the nitrate reducers. The nitrate reducers are predominantly heterotrophic bacteria that possess a complex electron transport system(s) allowing the NO_3^- ion to serve anaerobically as a terminal acceptor of electrons.

$$\left(NO_3 \xrightarrow{2e} NO_2^-;\ \ NO_3^- \xrightarrow{5e} N_2;\text{or } NO_3^- \xrightarrow{Se} NH_3\right)$$

The organic compounds that serve as specific electron donors for these three known nitrate reduction processes are shown in Table. The nitrate reductase activity is common in bacteria and is routinely used in the simple nitrate reductase test to identify bacteria.

Table: Nitrate Reducers.

Physiologic Types of Nitrate Reductases	Electron Donor(s)	Representatives Organisms
Respiratory ($NO_3^- \rightarrow NO_2^-$)	Formate NADH	Escherichia coli Klebsiella aerogenes
Denitrifying	NADH	Pseudomonas aeruginosa
$NO_3^- \rightarrow N_2$	Pyruvate NADH, Succinate	Clostridium perfringens Paracoccus denitrificans
Assimilatory ($NO_3^- \rightarrow NH_3$	Lactate H2- formate NADH succinate NADH NADH lactate, glycerol-phosphate	Staphylococcus aureus Vibrio succinogenes Bacillus stearothermophilus Enterobacter aerogenes Escherichia coli

$$4AH_2 + HNO_3 \xrightarrow{\text{Nitrate reduction}} 4A + NH_3 + 3H_2O + energy$$

(AH2 = Organic subtrate which serves as electron donor)

A second group of anerobic respirers the sulfate reducers utilizes SO_4^2 ion in similar fashion.

$(SO_4^{2-\ 8e:} H_2S)$:

$$4AH_2 + H_2SO_4 \xrightarrow{\text{Sulfate reduction}} 4A + H_2S + 4H_2O + energy$$

The third group the fumarate respirers are anaerobic bacteria that require exogenousHOOC-CH=CH-COOH for growth. Fumarate is reduced to succinate (HOOC-CH2-COOH), which is secreted as a by- product.

$$AH_2 + HOOC = \overset{H}{\overset{\|}{C}} - \overset{H}{\overset{\|}{C}}OOH \xrightarrow{\text{Furmate reduction}} A + HOOC + CH_2 + CH_2 - COOH + energy$$

Organisms of still another of still specialized group of anaerobic respirers, the methanogens, produce methane gas CO_2 & CH_4 as a metabolic end product of microbial growth. H_2 gas is the growth substrate; CO_2 is the terminal electron acceptor.

$$4H_2 + CO_2 \xrightarrow{CO_2\text{ reduction}} CH_4 + 2H_2O + energy$$

The methanogens are among the most anaerobic bacteria known, being very sensitive to small concentrations of molecular O_2. They are also archaebacteria, which typically live in unusual and deleterious environments.

All of the above anaerobic respirers obtain chemical energy for growth by using these anaerobic energy-yielding oxidation reactions.

The Nitrogen Cycle

Nowhere can the total metabolic potential of bacteria and their diverse chemical-transforming capabilities be more fully appreciated than in the geochemical cycling of the element nitrogen. All the basic chemical elements (S, O, P, C, and H) required to sustain living organisms have geochemical cycles similar to the nitrogen cycle.

The nitrogen cycle is an ideal demonstration of the ecologic interdependence of bacteria, plants, and animals. Nitrogen is recycled when organisms use one form of nitrogen for growth and excrete another nitrogenous compound as a waste product. This waste product is in turn utilized by another type of organism as a growth or energy substrate. Figure shows the nitrogen cycle.

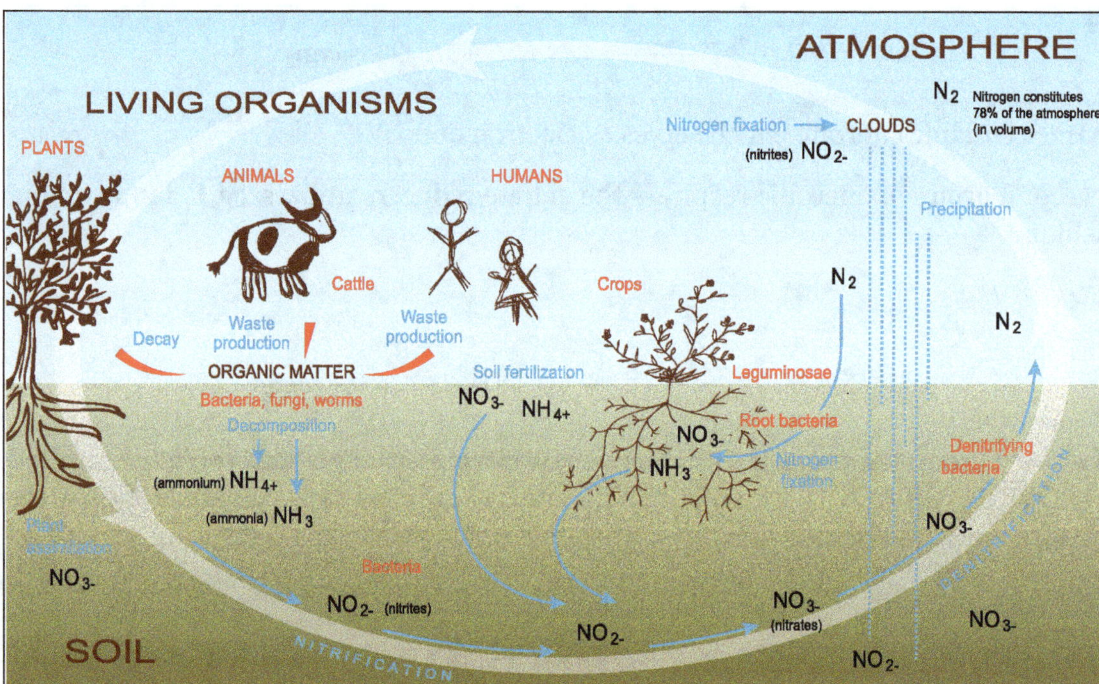

When the specific breakdown of organic nitrogenous compounds occurs, that is, when proteins are degraded to amino acids (proteolysis) and then to inorganic NH_3, by heterotrophic bacteria, the process is called ammonification. This is an essential step in the nitrogen cycle. At death, the organic constituents of the tissues and cells decompose biologically to inorganic constituents by a process called mineralization; these inorganic end products can then serve as nutrients for other life forms. The NH_3 liberated in turn serves as a utilizable nitrogen source for many other bacteria. The breakdown of feces and urine also occurs by ammonification.

The other important biologic processes in the nitrogen cycle include nitrification (the conversion of NH_3 to NO_3 by autotrophes in the soil; denitrification (the anaerobic conversion of NO_3 to N_2 gas) carried out by many heterotrophs); and nitrogen fixation (N_2 to NH_3, and cell protein). The latter is a very specialized prokaryotic process called diazotrophy, carried out by both free-living bacteria (such as Azotobacter, Derxia, Beijeringeia, and Azomona species) and symbionts (such as Rhizobium species) in conjunction with legume plants (such as soybeans, peas, clover, and bluebonnets). All plant life relies heavily on NO_3^- as a nitrogen source, and most animal life relies on plant life for nutrients.

Bacterial Classification

The classification of bacteria serves a variety of different functions. Because of this variety, bacteria may be grouped using many different typing schemes. The critical feature for all these classification systems is an organism identified by one individual (scientist, clinician, epidemiologist), is recognized as the same organism by another individual. At present the typing schemes used by clinicians and clinical microbiologists rely on phenotypic typing schemes. These schemes utilize the bacterial morphology and staining properties of the organism, as well as O_2 growth requirements of the species combined with a variety of biochemical tests. For clinicians, the environmental reservoir of the organism, the vectors and means of transmission of the pathogen are also of great importance.

Scientists interested in the evolution of microorganisms are more interested in taxonomic techniques that allow for the comparison of highly conserved genes among different species. As a result of these comparisons a phylogenetic tree can be developed that displays the degree of relatedness of different organisms. A relatively new application of this technology has been the recognition and characterization of noncultivatable pathogens and the diseases that they cause.

Phenotypic Classification Systems

There is a chart at the end of these lecture notes on the general phenotypic classification of many of the clinically important bacteria. This is provided as a reference. By the end of the course you will be able to recognize most of these microorganisms.

Gram stain and bacterial morphology: Of all the different classification systems, the Gram stain has withstood the test of time. Discovered by H.C. Gram in 1884 it remains an important and useful technique to this day. It allows a large proportion of clinically important bacteria to be classified as either Gram positive or negative based on their morphology and differential staining properties. Slides are sequentially stained with crystal violet, iodine, then destained with alcohol and counter-stained with safranin. Gram positive bacteria stain blue-purple and Gram negative bacteria stain red. The difference between the two groups is believed to be due to a much larger peptidoglycan (cell wall) in Gram positives. As a result the iodine and crystal violet precipitate in the thickened cell wall and are not eluted by alcohol in contrast with the Gram negatives where the crystal violet is readily eluted from the bacteria. As a result bacteria can be distinguished based on their morphology and staining properties.

Some bacteria such as mycobacteria (the cause of tuberculosis) are not reliably stained due to the large lipid content of the peptidoglycan. Alternative staining techniques (Kinyoun or acid fast stain) are therefore used that take advantage of the resistance to destaining after lengthier initial staining.

Growth Requirements: Microorganisms can be grouped on the basis of their need for oxygen to grow. Facultatively anaerobic bacteria can grow in high oxygen or low oxygen content and are among the more versatile bacteria. In contrast, strictly anaerobic bacteria grow only in conditions where there is minimal or no oxygen present in the environment. Bacteria such as bacteroides found in the large bowel are examples of anaerobes. Strict aerobes only grow in the presence of significant quantities of oxygen. Pseudomonas aeruginosa, an opportunistic pathogen, is an example of a strict aerobe. Microaerophilic bacteria grow under conditions of reduced oxygen and sometimes also require increased levels of carbon dioxide. Neisseria species (e.g., the cause of gonorrhea) are examples of micraerophilic bacteria.

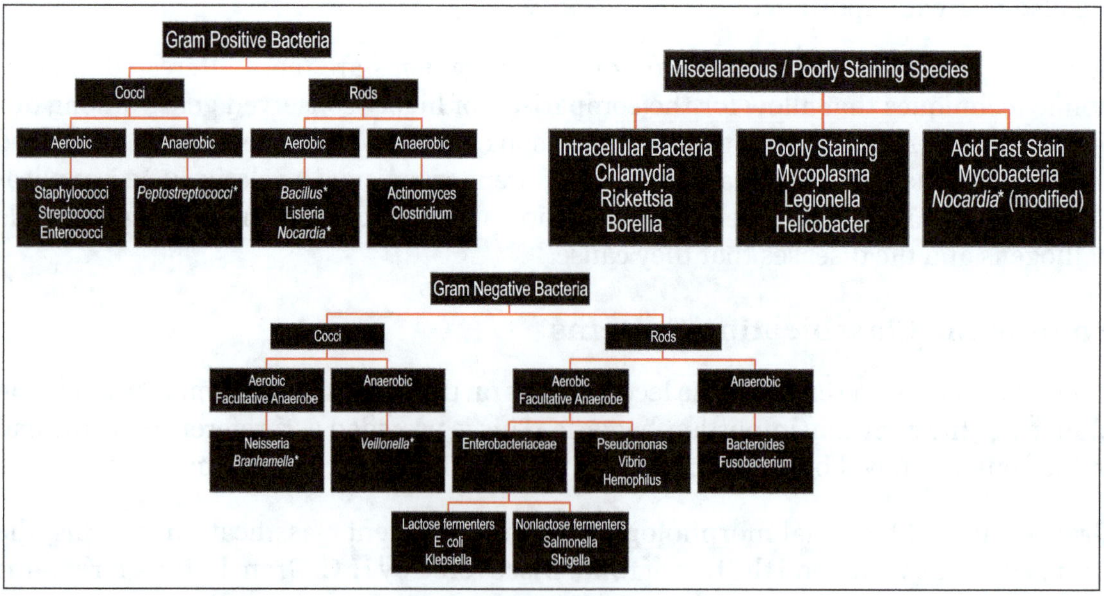

- Biochemical reactions: Clinical microbiology laboratories typically will identify a pathogen in a clinical sample, purify the microorganism by plating a single colony of the microorganism on a separate plate, and then perform a series of biochemical studies that will identify the bacterial species.
- Serologic systems: Selected antisera can be used to classify different bacterial species. This may be based on either carbohydrate or protein antigens from the bacterial cell wall or the capsular polysaccharide. Group A streptococcal M proteins or O and H polysaccharide antigens of salmonella.
- Environmental Reservoirs: When considering likely pathogens it is also important to know which of the different species are found in different locations. Environmental reservoirs are generally divided into those that are endogenous (i.e., on or within the human body) and exogenous (somewhere in the environment). When considering the likely cause of an infection the likely source of the infection is important in your differential diagnosis. For example, an anaerobic organism resident in the large bowel is the likely cause of an abdominal abscess that develops after large bowel surgery. A skin rash developing in a hiker with a history of multiple tick bites is more likely to be borrelia, the agent of Lyme disease. An outbreak of food poisoning traced to imported unpasteurized cheese might be due to listeria.

Endogenous reservoirs account for a large proportion of human infections. Many parts of the body have their own normal flora. S. epidermidis is found on the skin. Viridans streptococci are a part of the normal oropharyngeal flora and S. aureus is a commensal of the anterior nares.

Genotypic Systems

- Universal Phylogenetic Tree: Woese has developed a "universal phylogenetic tree" for all living organisms that establishes a tripartite division of all living organisms– bacteria, archaea and eucarya. His work is based on a comparison of 16s ribosomal RNA sequences. These sequences are highly conserved and undergo change at a slow, gradual and consistent rate. They are therefore useful for making comparisons among the different living organisms.
- Ribosomal RNA (rRNA) sequence analysis: This has emerged as a major method for classification. It has been used (as described above) to establish a phylogenetic tree. In addition, it is now also used to rapidly diagnose the pathogen responsible for an infection, to help select appropriate therapy and to identify noncultivatable microorganisms.
- Molecular subtyping: Sometimes it is necessary to determine whether strains from the same species are the same or different. For example, if there is an outbreak of infections that appear due to the same bacterial species, the hospital

epidemiologist will want to know if all of the infections are due to the same strain. Clues can be obtained by examining the biochemical studies or the antibiotic susceptibility profile, but a more reliable method is by molecular analysis. Pulsed Field Gel Electrophoresis (PFGE) is the most frequently used molecular technique. Chromosomal DNA is digested with a restriction enzyme that makes relatively infrequent cuts in the DNA and as a result creates large DNA fragments. The DNA fragments from the different strains are then run on a gel and compared.

Extremophiles

Extremophiles is an organism that is tolerant to environmental extremes and that has evolved to grow optimally under one or more of these extreme conditions, hence the suffix phile, meaning "one who loves."

Extremophile bacteria (living inside tube worms) that live on rocks near "black smoker" vents, such as the high-temperature, high-pressure Sully hydrothermal vent in the Main Endeavour Vent Field of the northeast Pacific Ocean, use chemosynthesis to harness chemical energy from toxic hydrogen sulfide gas released from the vent.

Extremophilic organisms are primarily prokaryotic (archaea and bacteria), with few eukaryotic examples. Extremophiles are defined by the environmental conditions in which they grow optimally. The organisms may be described as acidophilic (optimal growth between pH 1 and pH 5); alkaliphilic (optimal growth above pH 9); halophilic (optimal growth in environments with high concentrations of salt); thermophilic (optimal growth between 60 and 80 °C [140 and 176 °F]); hyperthermophilic (optimal growth above 80 °C [176 °F]); psychrophilic (optimal growth at 15 °C [60 °F] or lower, with a maximum tolerant temperature of 20 °C [68 °F] and minimal growth at or below 0 °C [32 °F]); piezophilic, or barophilic (optimal growth at high hydrostatic pressure); oligotrophic (growth in nutritionally limited environments); endolithic (growth within rock or within pores of mineral grains); and xerophilic (growth in dry conditions, with low water availability). Some extremophiles are adapted simultaneously to multiple stresses (polyextremophile); common examples include thermoacidophiles and haloalkaliphiles.

Extremophiles are of biotechnological interest, as they produce extremozymes, defined as enzymes that are functional under extreme conditions. Extremozymes are useful in industrial production procedures and research applications because of their ability to remain active under the severe conditions (e.g., high temperature, pressure, and pH) typically employed in these processes.

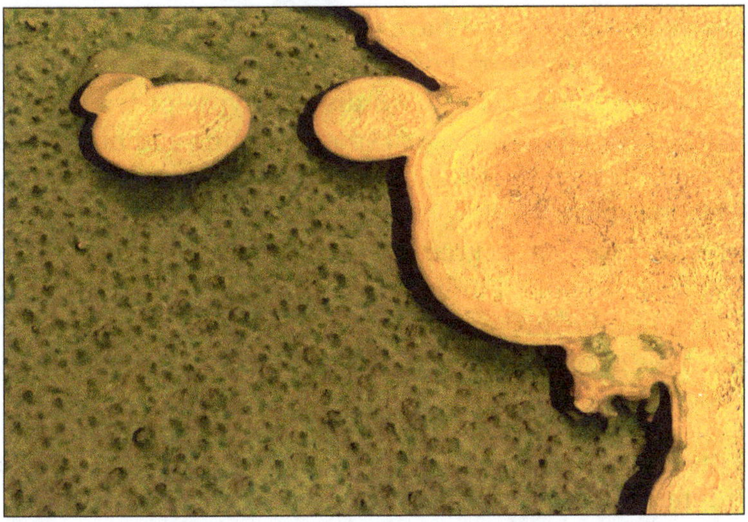

Figure: Extremophile; archaea: Yellow mats of extremophile archaea in a geothermal pool in Yellowstone National Park.

The study of extremophiles provides an understanding of the physicochemical parameters defining life on Earth and may provide insight into how life on Earth originated. The postulations that extreme environmental conditions existed on primitive Earth and that life arose in hot environments have led to the theory that extremophiles are vestiges of primordial organisms and thus are models of ancient life.

Extremophiles are also of research importance in the field of astrobiology. Extremophiles that are active at cold temperatures are of particular interest in this field, as the majority of the bodies in the solar system are frozen. The discovery of microorganisms with unusual biochemical properties, such as the ability to use arsenic rather than phosphorus for their growth, are also of interest to astrobiology, since extraterrestrial environments may favour life-forms that use or are built from elements not typically found in life on Earth. Thus, understanding the limits of life on Earth provides scientists with information about the possible existence of extraterrestrial life and provides clues about where and how to search for life on other solar bodies.

Thermophilic Bacteria

Thermophilic bacteria are those that grow at temperatures above the maximum temperature for the great majority of bacteria, especially the pathogenic forms. The maximum temperature for the pathogenic bacteria is about 45 °C. Their optimum temperature is about 37.5 °C. The true thermophiles show no growth, or only very feeble growth, below 40 to 45 °C. Their development requires temperatures above 50 °C and some are able to develop at a temperature of 80 °C though most abundant growth is shown at 60 to 70 °C.

A group of facultative thermophilic bacteria has been discovered which develop at room temperature, about 20 °C and have their optimum temperature at about 50 °C and their maximum temperature at about 60 °C. In this group belong some of the well-known spore-forming soil organisms. Two distinct types of true thermophilic bacteria have been found in the laboratory, viz., spore-forming and non-sporeforming rods, and thread forms belonging in the class of Nocardia. Most of the thermophilic bacteria studied belonged to the spore-forming class and these varied in size of rods, size and location of the spores, and to a less degree in their biological characters. The sources from which these bacteria are obtained indicate that this group is widely distributed in nature. They were found in the dust collected on apparatus in the laboratory, on clothes, in soil, in sewage, in the gastro-intestinal contents and feces of rabbits and guinea-pigs, on grain and in horse manure.

The isolation of these bacteria from the different materials is an easy matter, using the plate method with nutrient agar medium, the plates being incubated at a temperature above 50 °C. Under these conditions the colonies developing on the agar medium are certain to be thermophilic and can be obtained in pure culture by transplanting on to tubes of slanted agar.

The biological characters of the 60 cultures studied showed marked variation, especially with regard to their action on gelatin, milk, starch solution and nitrate solution. Eleven cultures failed to show any liquefaction of gelatin others showed only slight liquefaction, while others showed complete liquefaction in three or four days. The action on the gelatin was ascertained by cultivating at 51 °C for three or four days, and then placing the tubes at a low temperature to note whether liquefaction had occurred.

Most of the true thermophilic cultures showed only slight acidity in litmus milk, though several showed distinct acidity with coagulation but no digestion of the casein. Most of the facultative thermophilic cultures showed a primary acidity with coagulation followed by a slightly alkaline reaction and digestion of the casein. The action on starch varied in degree, as nearly all cultures appeared to have some action in splitting starch. A number of the cultures changed the starch to dextrin, maltose and glucose, though in many cultures no other change than the dextrin stage could be detected. The reduction of nitrate showed marked variation in the different cultures. Some showed only slight evidence of nitrite formation, others showed marked nitrite formation, while some of the cultures reduced the nitrate to ammonia. The true thermophiles belonging in the group of spore-forming bacteria differed in their morphologic and biologic character to such a degree that satisfactory classification was not possible. An attempt has been made to classify the different types of thermophilic bacteria and a comparison of the results obtained showed that amongst the more than sixty cultures studied, at least seven different rod forms and one thread form were encountered that may be classed as true thermophiles, while one type of rod was encountered that may be regarded as a facultative thermophile. The facultative thermophiles studied appeared to be more closely related to each other in that they formed short chains of rods with rounded ends having small oval central spores. These cultures all liquefied gelatin, fermented starch, and reduced nitrate to ammonia. Their action on milk showed at first a slight acidity followed by a slight alkalinity and slow peptonization of the casein. These cultures appeared to belong to a single species. The true thermophiles resisted a temperature of 100 °C, to an unusual degree varying from two to eight hours, so that it appears probable that the optimum temperature for growth is related to the heat resisting powers of the spores. The facultative thermophiles with their lower optimum temperature for growth also showed a lower resistance to the boiling temperature. This group of bacteria was killed in from one-fourth to one hour at 100 °C.

Two cultures of thread-form organisms were studied which had thermophilic properties. These two cultures represented two distinct varieties in their biological activities. On agar they showed a distinct white, mealy growth after several days and this fact has led to their classification with Nocardia. These two cultures differed in their action on milk and on starch solution, the one did not change milk but fermented starch to maltose, the other produced an acid reaction in milk with coagulation and subsequent digestion of the casein but only very slight action on starch. Both cultures liquefied gelatin but neither reduced nitrate. The maximum temperature of growth was about 70 °C, and both were killed at 100 °C, in two hours. Although the thermophilic bacteria are widely distributed in nature it is evident that their activities can be brought into play under exceptional conditions. These organisms have been found in thermal springs where temperature conditions favor their development. In the surface layers of the soil in the temperate and torrid zones the temperature frequently rises to a point where they can develop and it is probable that they find an important field for action under these conditions. There are other conditions in nature where it is probable that these organisms find suitable environment for their activities. In the fermentation of silage temperatures have been recorded as high as 55 °C but more commonly the temperature ranges between 35 and 45 °C. The production of heat is probably not due to chemical changes in the fodder, by the tissue cells, but primarily to the activities of the thermophilic bacteria.

Thermophilic Bacteria

	TRUE THERMOPHILES										Facultative
	Type 1	Type 2	Type 3	Type 4	Type 5		Type 6	Type 7	Type 8		Type 9
					Var. a	Var. b			Var. a	Var. b	
Source of organisms......	Dust and contaminated milk medium	Dust and soil	Dust, soil, horse manure	Dust, soil, horse manure	Dust, g-pig feces, horse manure	Dust, cheese, g-pig feces,	Horse manure	Horse manure	Rabbit's stomach	Contamination on agar	Dust, soil, horse manure
Form......	Rod	Rod	Rod	Rod	Rod	Rod	Rod	Rod	Threasd	Thread	Rod
Motility	+	+	+	+	+	+	+	+	-	-	+
Size in microns....	0.6-0.8x 3-4	0.5-0.7x 2-3	0.6-0.8x 4-5	0.6-0.8x 2-4	0.2-0.4x 2-3	0.2-0.4x 2-3	0.5-0.6x 2-2.5	0.3-0.4x 1.5-2	-	-	0.5-0.7x 2-3
Groupings...	Chains	Chains	Singly	Singly	Singly	Singly	Singly	Singly			Short chains
Location of spores...	Central	Central	Central	Polar	Polar	Polar	-	-	Conidia	Conidia	Central
Spores larger or smaller diameter or rods..	Smaller	Greater	Greater	Greater	Greater	Greater	-	-			Smaller
Gelatin liquefied	+	-	+	+	+	-	+	-	+	+	+
Nitrate reduced	NH_3	N_2O_2	±	N_2O_3	N_2O_3		±	N_2O_2	±	±	NH_2
Strach rediced	+	±	+	+	±	±	-	-	+	-	+
Action in litmus milk	Aid coagulation	Slight acidity	Slight acidity	Slight acidity	Slight acidity	Slight acidity	Slight acidity	Slight acidity	Slight acidity	Acid coagulation and digestion	Acid coagulation and digestion
Minimum temperature....	50-60	37	37-50	50	37	37	37	37	60	60	20-37

	Type 1	Type 2	Type 3	Type 4	Type 5		Type 6	Type 7	Type 8		Type 9
					Var. a	Var. b			Var. a	Var. b	
Source of organisms......	Dust and contaminated milk medium	Dust and soil	Dust, soil, horse manure	Dust, soil, horse manure	Dust, g-pig feces, horse manure	Dust, cheese, g-pig feces,	Horse manure	Horse manure	Rabbit's stomach	Contamination on agar	Dust, soil, horse manure
Form......	Rod	Rod	Rod	Rod	Rod	Rod	Rod	Rod	Threasd	Thread	Rod
Motility	+	+	+	+	+	+	+	+	-	-	+
Size in microns....	0.6-0.8x 3-4	0.5-0.7x 2-3	0.6-0.8x 4-5	0.6-0.8x 2-4	0.2-0.4x 2-3	0.2-0.4x 2-3	0.5-0.6x 2-2.5	0.3-0.4x 1.5-2	-	-	0.5-0.7x 2-3
Groupings...	Chains	Chains	Singly	Singly	Singly	Singly	Singly	Singly			Short chains
Location of spores...	Central	Central	Central	Polar	Polar	Polar	-	-	Conidia	Conidia	Central
Spores larger or smaller diameter or rods..	Smaller	Greater	Greater	Greater	Greater	Greater	-	-			Smaller
Gelatin liquefied	+	-	+	+	+	-	+	-	+	+	+
Nitrate reduced	NH_3	N_2O_2	±	N_2O_3	N_2O_3		±	N_2O_2	±	±	NH_2
Strach rediced	+	±	+	+	±	±	-	-	+	-	+
Action in litmus milk	Aid coagulation	Slight acidity	Slight acidity	Slight acidity	Slight acidity	Slight acidity	Slight acidity	Slight acidity	Slight acidity	Acid coagulation and digestion	Acid coagulation and digestion
Minimum temperature....	50-60	37	37-50	50	37	37	37	37	60	60	20-37
Maximum temperature....	75-80	70	70-75	75-80	60-70	60	60	60	70	70	50-60
Thermal-death point at 100 °C in minutes.	400	300	200	180	120	60	5	5	120	120	15-60

Under these conditions it seems likely that the facultative thermophiles begin the fermentation processes and when the temperature rises to a sufficient degree then the true thermophiles begin to develop. Their ability to act on carbohydrates and protein indicates that they probably find sufficient food for development in various fodder crops. The spoiling of hay and other fodder crops and of grain, when insufficiently cured, is no doubt due largely to the action of the thermophilic bacteria. The charring effect noticed in piles of improperly cured hay and fodder and in manure is probably caused by the thermophilic bacteria. The conditions of temperature under which the thermophilic,bacteria thrive are paradoxical when we note that the optimum temperature of growth is at about the temperature at which ordinary egg and serum albumin begin to coagulate. The maximum temperature of growth is decidedly above the temperature at which egg and serum albumin coagulate. This phenomenal characteristic may be due to the reaction of the medium in which the bacteria are growing or to the mineral content of their own protoplasm.

Mesophilic Bacteria

Mesophiles are microorganisms such as some species of Bacteria, Fungi, and even some Archaea that are best active at median temperatures. For instance, bacterial species involved in biodegradation (i.e., digestion and decomposition of organic matter), which are more active in temperatures ranging from approximately 70° - 90 °F (approx. 15°–40 °C), are termed mesophilic bacteria. They take part in the web of micro-organic activity that forms the humus layer in forests and other fertile soils, by decomposing both vegetable and animal matter.

At the beginning of the decomposition process, another group of bacteria, psychrophylic bacteria, start the process because they are active in lower temperatures up to 55 °F (from below zero up to 20 °C), and generate heat in the process. When the temperature inside the decomposing layer reaches 50–100 °F, it attracts mesophilic bacteria to continue the biodegradation. The peak of reproductive and activity of mesophilic bacteria is reached between 86–99 °F (30–37 °C), and further increases the temperature in the soil environment. Between 104–170 °F (40–85 °C, or even higher), another group of bacteria (thermophyllic bacteria) takes up the process that will eventually result in organic soil, or humus. Several species of fungi also take part in each decomposing step.

Mesophilic bacteria are also involved in food contamination and degradation, such as in bread, grains, dairies, and meats. Examples of common mesophilic bacteria are Listeria monocytogenes, Pesudomonas maltophilia, Thiobacillus novellus, Staphylococcus aureus, Streptococcus pyrogenes, Streptococcus pneumoniae, Escherichia coli, and Clostridium kluyveri. Bacterial infections in humans are mostly caused by mesophilic bacteria that find their optimum growth temperature around 37 °C (98.6 °F), the normal human body temperature. Beneficial bacteria found in human intestinal flora are also mesophiles, such as dietary Lactobacillus acidophilus.

Piezophiles

Piezophiles (barophiles) are organisms whose survival and reproduction is optimized to high pressures, such as those in deep sea environments. Piezophiles are found primarily in the depths of the ocean, which has an average pressure of 38 MPa (megapascals) and reaches 110 MPa at its deepest point in the Marianas Trench, in comparison to a pressure of 0.1 MPa at sea level. Remarkably, the most pressure resistant organism currently known, Pyrococcus yayanosii, can survive at pressures up to 150MPa. In order to inhabit such extreme environments, these organisms have evolved various mechanisms to counteract the effects of the elevated pressures they endure. While relatively little is known about piezophiles as of yet, preliminary research indicates potential industrial and biotechnological applications for high pressure adaptation mechanisms.

Phylogenetic analysis of the 16S rRNA sequences of piezophilic bacteria shows a high level of similarity to non-piezophilic bacteria from Antarctica, and all but one of the currently known piezophilic species are also psychropiezophiles. Because of this, it is theorized that piezophily arose from psychrophiles, whose pre-existing adaptations to extreme cold overlap significantly with the adaptations necessary to survive in high pressure environments. The vast majority of piezophiles are bacteria, and can be grouped into one of five genera – Shewanella, Colwellia, Photobacterium, Moritella, and Psychromonas. The first example of a piezophilic archaeon is the recently discovered thermopiezophile Pyrococcus yayanosii.

Piezophiles	Exhibit optimal reproduction rates at hydrostatic pressures exceeding 10MPa, found throughout the deep-biosphere.
Psychropiezophiles	All known bacterial piezophiles fall into this category. Exhibit optimal reproduction rates at hydrostatic pressures exceeding 10MPa and at temperatures ranging from 2-4 °C found throughout the deep-biosphere.
Thermopiezophiles	The only current example is the archaeon pyrococcus vavanosii. Exhibits optimal reproduction rates at MPa and 98 °C.

Physical Environment

Piezophilic organisms are found in what is called the deep-biosphere, defined as those environments more than 1000m below sea level and with hydrostatic pressures greater than 10MPa. The deep-biosphere is for the most part oligotrophic, characterized by a lack of sunlight and nutrients – less than 1% of available organic material descends to the levels inhabited by these organisms. The piezophiles that have been isolated as of yet can be divided into thermopiezophiles and psychropiezophiles. The only known thermopiezophile is Pyrococcus yayanosii strain CH1, which is found in the vicinity of hydrothermal vents – fissures in the sea floor from which water geothermally heated to temperatures up to 400 °C is emitted, resulting in eutrophic, microbially dense communities. Psychropiezophiles are also found in the depths of the ocean, but in areas not heated by geothermal energy. The average temperature in this environment is approximately 2 °C.

Figure: Hydrothermal vents on the sea floor have the potential to sustain many thermophilic organisms, including Pyrococcus yayanosii.

Adaptation to High Pressure

High pressure has a similar effect on cell membranes as low temperature, which reduces fluidity by increasing packing of the fatty acyl chains of phospholipids. Most piezophiles are subject to both intense hydrostatic pressure and low temperatures, and the decrease in membrane fluidity this causes can result in a gel-like membrane, which interferes with nutrient uptake and cell signalling mechanisms. In piezophiles, this is prevented by increasing the proportion of fluidizing monounsaturated and polyunsaturated fatty acids in their membranes, which cannot be packed as tightly as saturated fatty acids.

High hydrostatic pressure is also known to change the conformation of cellular proteins. While pressures in the range relevant to piezophiles are not sufficient to denature protein, the small conformational changes they cause have been found to change or inhibit protein function. Amino acid sequencing of proteins from piezophilic Shewanella bacteria indicates lower concentrations of proline residues, whose cyclic side chains disrupt alpha helices, as well as glycine residues, whose small side chains with high conformational flexibility also destabilize helical protein structure. These modifications lead to an overall decrease in protein flexibility, and increased stability in high pressure environments. At the gene expression level, high hydrostatic pressure has been shown to up-regulate the transcription of genes encoding chaperone proteins, which maintain proper folding, and therefore function of the proteins they act on.

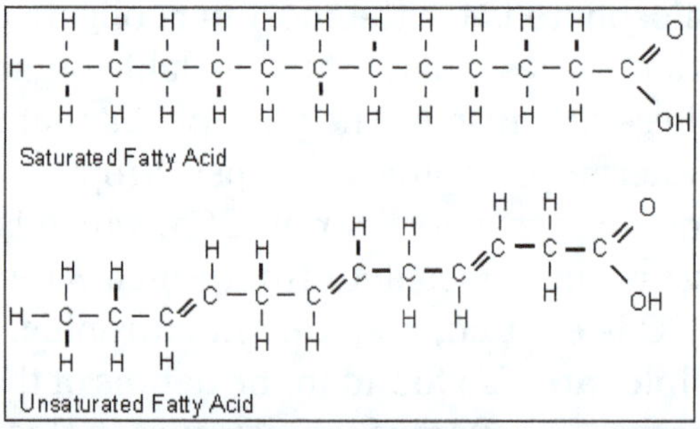

Figure: The double-bonded "kinks" in unsaturated fatty acids prevent them from being packed as tightly as saturated fatty acids.

Possible Applications of High Pressure Adaptation Mechanisms

Research on piezophiles has yielded promising results in regulating gene expression with hydrostatic pressure. This regulation mechanism was first discovered in Photobacterium profundum, in which hydrostatic pressure was found to act on the ToxR-ToxS complex, which in turn controls expression of two outer membrane proteins, OmpH and OmpL. At the bacterium's optimal pressure of 28 MPa, OmpH is dominant, but when exposed to decreasing hydrostatic pressure the ratio of OmpH to OmpL decreased accordingly. A promoter has also been found in Shewanella benthica and Shewanella violacea which increases expression of its downstream genes when grown at a pressure of 50 MPa. It has been theorized that inserting specific genes downstream of this promoter via homologous recombination, or insertion of the promoter sequence upstream of a desired gene would allow pressure regulated control of gene expression.

A number of proteins are in fact more stable at high hydrostatic pressures, and increasing pressure has been shown to prevent denaturation of some proteins even past their inactivation temperatures. For example, the half-life of Taq polymerase, an enzyme essential in the polymerase chain reaction process, was found to increase from 5 min at 3 MPa to 12 min at 45 MPa. This effect has also been observed for other DNA polymerases, as well as various hydrogenases (which catalyse oxidation of H_2), and α-glucosidases (which are involved in the breakdown of starch and glycogen). The study of these proteins and why they stabilize under these conditions is of use in the design of synthetic proteins to increase efficiency of processes that can be performed at high hydrostatic pressures, such as food processing and antibiotic production.

Psychrophilic Bacteria

Psychrophilic ("cold loving") microorganisms, particularly bacteria, have a preferential temperature for growth at less than 59 °Fahrenheit (15 °Celsius). Bacteria that can grow at such cold temperatures, but which prefer a high growth temperature, are known as psychrotrophs.

The discovery of psychrophilic microorganisms and the increasing understanding of their functioning have increased the awareness of the diversity of microbial life on Earth. So far, more than 100 varieties of psychrophilic bacteria have been isolated from the deep sea. This environment is very cold and tends not to fluctuate in temperature. Psychrophilic bacteria are abundant in the near-freezing waters of the Arctic and the Antarctic. Indeed, in Antarctica, bacteria have been isolated from permanently ice-covered lakes. Other environments where psychrophilic bacteria have been include high altitude cloud droplets.

Psychrophilic bacteria are truly adapted for life at cold temperatures. The enzymes of the bacteria are structurally unstable and fail to operate properly even at room (or ambient) temperature. Furthermore, the membranes of psychrophilic bacteria contain much more of a certain kind of lipid than is found in other types of bacteria. The lipid tends to be more pliable at lower temperature, much like margarine is more pliable than butter at refrigeration temperatures. The increased fluidity of the membrane makes possible the chemical reactions that would otherwise stop if the membrane were semi-frozen. Some psychrophiles, particularly those from the Antarctic, have been found to contain polyunsaturated fatty acids, which generally do not occur in prokaryotes. At room temperature, the membrane of such bacteria would be so fluid that the bacterium would die.

Aside from their ecological curiosity, psychrophilic bacteria have practical value. Harnessing the enzymes of these organisms allows functions such as the cleaning of clothes in cold water to be performed. Furthermore, in the Arctic and Antarctic ecosystems, the bacteria form an important part of the food chain that supports the lives of more complex creatures. In addition, some species of psychrophiles, including Listeria monocytogenes are capable of growth at refrigeration temperatures. Thus, spoilage of contaminated food can occur, which can lead to disease if the food is eaten. Listeriosis, a form of meningitis that occurs in humans, is a serious health threat, especially to those whose immune system is either not matures or is defective due to disease or therapeutic efforts. Other examples of such disease-causing bacteria include Aeromonas hydrophila, Clostridium botulinum, and Yersinia enterocolitica.

Gram-positive Bacteria

These are bacteria whose classification has been based on their ability to retain the crystal violet dye after a brief wash with alcohol in the differential gram staining. Gram-positive bacteria stain purple while the Gram-negative bacteria stain Pink, after losing the purple color during the alcohol was thus taking up the safranin.

Gram staining has been especially used because of its ability to differentiate bacteria base on their cell wall content, a major characteristic that classifies bacteria into two types namely, Gram-positive and Gram-negative bacteria.

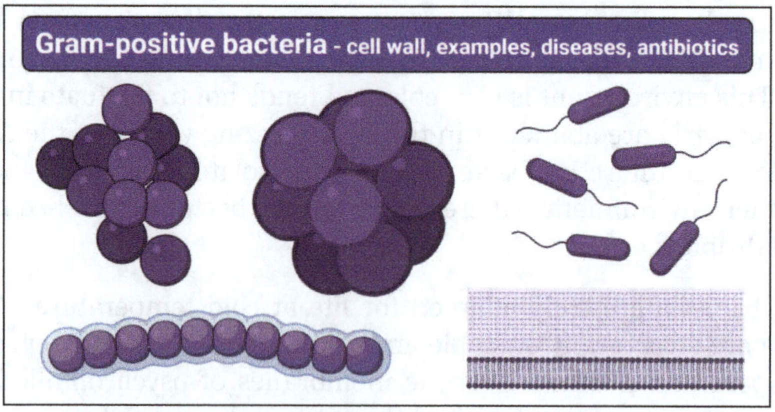

This classification was based majorly on the components they possess on their cell wall which enables them to stain differently. The bacterial cell is made up of Mucin, Peptioglycan, and mucopeptides, which majorly function to give the cell wall its rigid strength.

For the gram-positive cell wall, it has a thickness of about 20-80nm thickness made up of a thick peptidoglycan layer outside its cell membrane, unlike the thin layer of gram-negative bacteria (10-15nm) which has a very thin layer of the peptidoglycan of 2-7nm but has a thicker lipid layer making it quite complex than the Gram-positive cell wall.

Gram-positive Bacteria Characteristics

These bacteria have very distinct features that characterize it and differentiate it from other types of bacteria. These include:

- They lack an outer membrane.
- They have a thin layer of the cytoplasmic lipid layer.
- They have a thick peptidoglycan layer.
- The peptidoglycan layer has a large quantity of teichoic acid and a thin lipid layer, made up of lipoteichoic acid which plays a major role in bacterial adherence.
- The peptidoglycan layer plays a key role in maintaining the rigidity of the cell wall by crosslinking by the assistance of the DD-transpeptidase.
- They also have a thin layer of periplasm, as compared to that in the Gram-negative bacteria.
- Some have a locomotive apparatus, a flagellum that has two basal bodies for support unlike gram-negative which has four basal bodies.
- Some have a strong capsule made up of polysaccharides.

Gram-positive Bacteria Shape

Despite most bacteria being differentiated by the Gram staining dyes, the observation under the Microscope reveals more features that can be used to define and characterize these bacteria.

Gram-positive by definition in shape can be classified into two:

- Cocci: Singular form is known as coccus, which is round or oval shape bacteria measuring 0.5-1.0 um in diameter. They occur in pairs or in chains or clusters or as singles. For example Staphylococcus app, Streptococcus app. Special spaces of cocci include tetrads, Sarcines.

- Bacilli: Also rod-shaped; the singular form is known as Bacillus. They are stick-like bacteria with a round tapered, square or swollen ends. They have a size of 1-10um in length and 0.3-1.0um in width. For example the Bacillus spp.

Other special shapes formed by Gram-positive bacteria include:

- Tetrad: A type of cocci shape occurring in square-clusters of fours, for example, Micrococcus spp.
- Sarcina (Octae): Thick-walled cocci shapes, occurring in clusters of four, or cubes of eight, for example, Sarcina app.

Gram-positive Bacteria Cell Wall

The thick Gram-positive bacterial cell is made up of a large quantity of peptidoglycan, teichoic acid, a thin lipid layer below the peptidoglycan layer and glycerol polymers.

Peptidoglycan

It is also known as murein, making up 90% of the bacterial cell wall content. Its major role is to provide shape and maintain cell wall strength and rigidity. It is a high-quality polymer made up of two identical sugar derivates, named N-acetylglucosamine and N-acetylmuramic acid and a chain of L-amino acids and three distinct D-amino acids that are rarely found in proteins i.e D-glutamic acid, D-alanine, and meso-diaminopimelic, which protect the cell wall from attack by peptidase enzymes. The D-amino acids and the L-amino acids connect to the N-acetylmuramic acid, L-amino acid specifically the L-lysine can replace the meso-diaminopimelic acid.

This interconnection of peptidoglycan subunit makes the peptidoglycan Strong to maintain the bacterial shape and integrity, with the ability to be elastic and stretch. The peptidoglycan is also permeable allowing molecules to move in and out of the bacterial cell.

Teichoic Acid

This is a fortified wall made up copolymers of glycerol. It is water-soluble making up to 50% of the total dry weight of the bacterial cell wall. It is either directly connected to the peptidoglycan, covalently or to the cell membrane (lipoteichoic acid). The direct link to the peptidoglycan in by the 6-hydroxyl N-acetylmuramic acid.

It is negatively charged and they extend to the peptidoglycan surface, giving the bacterial cell wall a negative charge. It also contributes to maintaining the structure of the cell wall. It is completely absent in gram-negative bacteria.

Lipid

They have a thin layer of lipids below the peptidoglycan, of about 2-5%, which functions to anchor the bacterial cell wall.

Gram-positive Bacteria Examples and Diseases

Here we describe various Gram-positive bacteria, their basic morphological features the diseases they cause in Humans:

Gram-Positive Bacteria	Bacterial infection: Clinical Features
Staphylococcus aureus	• They cause localized and deep localized skin infections such as Furuncles, carbuncles, Wound infection, Cellulitis, Impetigo. • Respiratory infections such as pneumonia. • Cardiovascular infections such as Endocarditis and Septicemia. • Musculoskeletal Infections such as Osteomyelitis and arthritis.
Streptococcus pyogenes	• Acute pharyngitis or pharyngotonsillitis. • Impetigo. • Erysipelas. • Puerperal sepsis. • Invasive group A streptococcal disease (Necrotizing fasciitis/ myositis). • Streptococcal toxic shock syndrome. • Acute rheumatic fever. • Acute glomerulonephritis.
Streptococcus pneumoniae	• Community-acquired Acute bacterial pneumonia. • Adult bacterial meningitis. • Otitis media in children. • Bacteremia/sepsis. • Meningitis.
Bacillus anthracis	• It causes Anthrax disease of the pulmonary system (woolsorter's disease) associated with hemorrhagic necrosis and edema of the Mediastinum, with coughing, sepsis which spreads to the gastrointestinal tract causing bowel ulceration and hemorrhagic meningitis. • The anthrax disease of the cutaneous layer with pruritic papule lesions and edema, lymphadenopathy, Lymphangitis, systemic symptoms of fever, malaise and headache.
Corneybacterium diphtheriae	• Diphtheria, a localized infection of the upper respiratory tract. • Cardiac conduction defects and myocarditis may lead to congestive heart failure and permanent heart damage if the disease extends to the lower tract. • Neuritis of cranial nerves and paralysis of eye and leg muscles. • Cutaneous diphtheria in wounded or cut skin causing chronic ulcerated nonhealing wounds.

Clostridium botulinum	• Causes botulism poisoning associated with visual disturbances (incoordination of eye muscles, double vision), inability to swallow, and speech difficulty; bulbar paralysis, respiratory paralysis, and cardiac arrest may lead to death.
Clostridium tetani	• Tetanus characterized by tonic contraction of voluntary muscles. Muscle spasm on the area of injury; infection of the jaw muscles which makes it difficult to open the mouth. • Tonic spasm of other body muscles and tetanic generalized muscle spasm.
Clostridium difficile	• Pseudomembranous Colitis associated with watery, bloody diarrhea, microabscess in the bowel.
Enterococcus faecium and Enterococcus faecalis	• They rarely caused disease in healthy persons. They cause disease in immune-compromised individuals with wounds and attached instruments like respirators, catheters causing urinary tract infections, bacteremia/sepsis, endocarditis, biliary tract infection, or intra-abdominal abscesses.
Listeria monocytogenes	• Listeriosis associated with septicemia and meningitis. • Focal and granulomatous skin lesions. • An asymptomatic vaginal disease spreads to newborns causing flu-like illness and meningitis in the newborn.

Gram-positive Bacteria Antibiotics

Gram-positive bacteria are known to cause several infections which may be disastrous to humans if not treated and managed on time and properly. For this reason, scientists manufactured chemotherapeutic agents, known as antibiotics which act against the bacterial agent causing the disease, eliminating it from the system by killing it. The mechanisms of elimination of the antibiotics vary.

Gram-Positive antimicrobial agents include:

Antimicrobial agent	Mechanism of action	Examples of Bacteria
ß-lactamases: Amoxicillin, methicillin, Oxacillin, Ampicillin, Penicillin G	Disruption of bacterial cell wall.	Staphylococcus aureus Streptococcus pneumonie Streptococcus pyogenes Corynebacteriun diphtheria Bcillus anthracis Clostridium botulinum
Vancomycin, Erythromycin, Azithromycin	Inhibits cell wall synthesis by preventing the crosslinking of the peptidoglycan peptidases.	Staphylococcus spp Streptococcus spp Bacillus spp Clostridium difficile

Bacitracin	Inhibits cell wall synthesis by preventing movement of the cytoplasmic membrane and peptidoglycan precursors.	Corneybacterium spp Bacillus anthracis
Macrolides: Azithromycin, Clarithromycin	Inhibit bacterial protein synthesis by preventing polypeptide elongation of 50s ribosomes.	Streptococcus pyogenes
Cephalosporin	Inhibition of cell wall synthesis (Disruption of peptidoglycan synthesis).	Streptococcus pneumoniae Bacillus anthracis
Aminoglycosides: Gentamicin	Inhibition of bacterial protein synthesis by the production of aberrant peptide chains at the 30s ribosomes.	Staphylococcus aureus Streptococcus pneumoniae Streptococcus pyogenes Enterococcus spp
Oxazolidinone	Inhibition of protein synthesis at the 50s ribosomes.	Enterococcus spp
Rifampicin	It inhibits the synthesis of nucleic acids by preventing the transcription of binding DNA-dependent RNA polymerase.	Bacillus anthracis
Sulfonamides: sulfamethoxazole	An antimetabolite that inhibits dihydropteroate synthase and disrupts the folic acid synthesis.	Listeria monocytogenes
Trimethoprim	An antimetabolite that inhibits dihydrofolate reductase and disrupts the folic acid synthesis.	Listeria monocytogenes

Staphylococcus Aureus

Staphylococcus aureus is a Gram-positive bacterium and causative agent of wide range of infectious diseases such as skin infections, bacteremia, endocarditis, pneumonia and food poisoning. The organism was originally a leading nosocomial pathogen and afterwards epidemiologically distinct clones emerged in community settings. S. aureus expresses number of virulence factors which help to establish infection by facilitating tissue attachment, tissue invasion and evading from host immune response. The ability to acquire resistance to multiple antibiotics classes makes S. aureus, a challenging pathogen to treat. Emergence and spread of S. aureus strains which are resistant to methicillin, referred to as methicillin-resistant S. aureus (MRSA) resulted in high morbidity, high mortality and increased treatment costs. Vancomycin remained gold standard drug to tackle these strains for years but the emergence of resistance restricted its clinical utility. Newer anti-MRSA antibiotics which were approved by U.S. FDA came as respite for clinicians. However, new antibiotic discovery efforts and non- antibiotic approaches to tackle MRSA should not be diminished considering the ability of the pathogen to acquire resistance to newer drugs quickly after their introduction in clinics.

Microscopic Morphology

S. aureus cells are Gram-positive and appear in spherical shape. They are often in clusters resembling bunch of grapes when observed under light microscope after Gram staining. The name 'Staphylococcus' was derived from Greek, meaning bunch of grapes (staphyle) and berry (kokkos). The scanning electron microscopic observation reveals roughly spherical shaped cells with smooth surface. The diameter of the cells ranges from 0.5 to 1.0 μM. The transmission electron microscopy of cells shows thick cells wall, distinctive cytoplasmic membrane and amorphous cytoplasm.

General Cultural and Biochemical Characteristics

S. aureus is an aerobic and facultative anaerobic organism that forms fairly large yellow or white colonies on nutrient rich agar media. The yellow colour of the colonies is imparted by carotenoids produced by the organism. The term 'aureus' is derived from Latin, which refers to the colour of gold. The organism is often haemolytic in blood agar due to production of four types of haemolysins (alpha, beta, gamma and delta). Nearly all isolates of S. aureus produce coagulase enzyme, a virulence factor that also helps in identification of the organism. The organism is salt tolerant, which is able to grow in mannitol-salt agar medium containing 7.5% sodium chloride. The organism is catalase positive and oxidase negative.

General Pathogenesis and Clinical Diseases

Pathogenesis

The process of S. aureus infections involves five stages. They are colonization, local infection, systemic dissemination and/or sepsis, metastatic infections and toxinosis. The organism is in carrier state in the anterior nares and can remain so without causing infections for weeks or months. The colonization proceeds to infection under certain predisposing factors such as prolonged hospitalization, immune suppression, surgeries, use of invasive medical devices and chronic metabolic diseases. Localized skin abscess develop when the organism is inoculated into the skin from a site of carriage. This can further spread and results in various clinical manifestations of localized infections such as carbuncle, cellulitis, and impetigo bullosa or wound infection. The organism can enter into blood and spread systemically to different organs causing sepsis. This haematogenous spread may result in endocarditis, osteomyelitis, renal carbuncle, septic arthritis and epidural abscess. Without a blood stream infection, specific syndromes can occur due to extra cellular toxins of S. aureus. These are toxic shock syndrome, scalded skin syndrome and foot borne gastroenteritis.

Hospital and Community Infections

S. aureus causes wide range of infections in human. The clinical infections of S. aureus are classified into community and nosocomial categories based on origin of infection.

These two types are distinct in clinical manifestations of the infections, antibiotic susceptibility and the genetic background of the infecting S. aureus strains. For decades, S. aureus has been predominately a nosocomial pathogen and is a leading cause of mortality and morbidity in hospitals. However, the community S. aureus infections are in rise. The important clinical S. aureus infections are bacteraemia, infective endocarditis, skin and soft tissue infections, osteoarticular infections and pleuropulmonary infections. Other clinical infections are epidural abscess, meningitis, toxic shock syndrome and urinary tract infections.

Virulence Factors

S. aureus possess battery of virulence factors. These factors enable the organism to be successful as pathogen that causes wide range of human and animal infections. Virulence factors help in attachment to host cells, breaking down the host immune shield, tissue invasion, causing sepsis and elicit toxin-mediated syndromes. This is the basis for persistent staphylococcal infections without strong host immune response. Based on their mechanism of action and role in pathogenesis, staphylococcal virulence factors are classified as represented in Table.

Table: Virulence factors of S. aureus and its characteristics.

Factors	Characteristics
Helping attachment to host tissues	
Microbial Surface Components Recognizing adhesive matrix molecules (MSCRAMM)	Cell surface proteins which interact with host molecules such as collagen, fibronectin & fibrinogen, thus, facilitate the tissue attachment. Staphylococcal protein A, fibronectin-binding proteins A and B, collagen-binding protein & clumping factor A & B belong to this family. They are also involved in host immune evasion.
Breaking/evading the host immunity	
Polysaccharide microcapsule	Resist the phagocytosis & killing by polymorphonuclear phagocyte.
Protein A	It binds to Fc portion of immunoglobulin, prevents opsonization, functions as super antigen & limits the host immune response.
Panton-Valentine leukocidin (PVL)	PVL is found in most of community-associated MRSA (CA-MRSA). PVL belongs to group of membrane pores forming proteins. It consists of two protein components (LukS-PV and LukF-PV) which act together as subunits and form porins on cell membrane of host cells, leading to leakage of cell contents and cell death.
Alpha-toxin (Alpha hemolysin)	It was the first bacterial exotoxin to be identified as a cell membrane pore former which causes cell leakage & death.
Chemotaxis-inhibitory protein of S. aureus (CHIPS)	CHIPS is an extracellular protein which inhibits the chemotaxis functioning of neutrophil and monocytes.

Tissue invasion	
Extracellular adherence protein (Eap)	An exoprotein which binds to host cell matrix, plasma proteins & endothelial cell adhesion molecule ICAM-1. In addition to the roles of adhesion and invasion, it also has immune-modulatory activity.
Proteases, lipases, nucleases, hyaluronatelyase, phospholipase C, metalloproteases (elastase), & Staphylokinase	These extracellular enzymes cause tissue destruction and, thereby, help in bacterial penetration into tissues.
Induces toxinosis	
Enterotoxins	S. aureus produces battery of enterotoxins which are potent gastrointestinal exotoxins. The Staphylococcal food poisoning is an intoxication which results from consumption of foods containing sufficient amount of preformed enterotoxins.
Toxic shock syndrome toxin -1 (TSST-1)	TSST-1 & some of enterotoxins are called as pyrogenic toxin super antigens. TSST-1 causes toxic shock syndrome especially in menstrual women.
Exfoliative toxins A and B	Serine proteases which selectively recognize and hydrolyze desmosomal proteins in the skin. ETs cause staphylococca-scalded skin syndrome, a disease predominantly affecting infants.

Antibiotic Resistance

Penicillin Resistance

The first beta-lactam antibiotic penicillin G was discovered in 1928 by Alexander Fleming and the drug was used in human as chemotherapeutic agent in 1941. The antibiotic was potent against Gram positive pathogens and a power weapon against Staphylococcal infections. However, first reports of S. aureus strains that were resistant to penicillin appeared after a year of its clinical use. Such penicillin-resistant isolates carried a plasmid gene, blaZ which encoded a beta-lactamase enzyme, referred to as penicillinase. The enzyme is capable of cleaving the beta-lactam ring of penicillin resulting inactivation of the antibiotic.

The emergence and spread of penicillinase-mediated resistance in S. aureus is referred to as first wave of resistance. This has spread in alarm proportions and became pandemic in the 1960s. About 80% of both community and hospital acquired S. aureus isolates were resistant to penicillin by late 1960s. By early 2000s, more than 90% of Staphylococcal isolates produced penicillinase enzyme irrespective of their community or hospital origin.

Methicillin Resistance

The penicillinase resistance in S. aureus was countered by the discovery of methicillin, penicillinase-stable semisynthetic penicillin. The drug was introduced into clinics in 1961 and subsequently strains showing methicillin resistance (MRSA) was reported in the same year. After the initial report, MRSA clones spread rapidly across the world but restricted to nosocomial settings. This is referred to as second wave of beta-lactam resistance in S. aureus. The methicillin resistance was mediated by the presence of mecA gene. The therapeutic outcome of MRSA infections was worse than methicillin sensitive S. aureus (MSSA) due to the underlying comorbid factors such as old age, immune suppression and, importantly, lack of effective antibiotics to treat MRSA, which were often multi-drug resistant. The rise in MRSA infections in hospitals resulted in high morbidity and mortality and increase in cost of health care.

The third wave of beta-lactam resistance in S. aureus began with reports of MRSA infections in community in early 1990s. These strains were phenotypically and genetically distinct from MRSA isolates from hospitalized patients, resulting in definitions of HA-MRSA and CA-MRSA. In the last decade, community MRSA strains invaded the hospital settings and the difference between HA and CA MRSA is now blurred.

Quinolone Resistance

Nalidixic acid, the prototype quinolone and the second generation quinolones (e.g. ciprofloxacin and norfloxacin) are predominately active towards Gram negative bacteria while third generation (e.g. levofloxacin) and fourth generation (e.g. moxifloxacin, gemifloxacin) quinolones exhibited improved and greater activity against Gram-positive bacteria. Quinolones exert their antibacterial action by inhibiting bacterial topoisomerases (topoisomerase IV and DNA Gyrase), which are essential for relieving DNA super coiling and separation of concatenated DNA strands. The resistance to quinolones in S. aureus arises in stepwise manner, due to point mutations primarily in GrlA subunit of topoisomerase IV and GyrA subunit of Gyrase. Additional mechanism by which S. aureus become resistant to quinolones is by expression of NorA efflux pumps.

The quinolone resistance in S. aureus is mostly associated with methicillin resistance though the mechanism of resistance and encoding genes are altogether different from each other. This could be due to higher usage of quinolones in hospital settings where the HA-MRSA prevalence is high resulting in selection of quinolone resistance. In year 2008, the fluoroquinolone resistance among MRSA isolates implicated in acute bacterial skin and skin structure infections (ABSSSIs) in hospitals was at 70.3%. Due to such high level of quinolone resistance among MRSA in hospital settings, even third- and fourth-generation quinolones have not been considered for treatment of MRSA. With respect to CA-MRSA, though they were susceptible to non-beta-lactam antibiotics including quinolones, the scenario has changed in recent years, with the rise in incidence of CA-MRSA infections which were multi-drug resistant.

Vancomycin Resistance

Vancomycin, a glycopeptide antibiotic, was discovered from a microbial source (Streptomycesorientalis) in 1952. The drug was approved for clinical use in 1958; however, it was eclipsed by methicillin and other anti-staphylococcal penicillins which were considered less toxic than vancomycin and equally efficacious against penicillin-resistant Staphylococci. Beginning early 1980s, there was sudden increase in vancomycin usage due to rise in HA-MRSA infections and emergence of pseudomembranous enterocolitis cause by Clostridium difficile in hospitalized patients. Clinical efficacy of vancomycin efficacy in treatment of MRSA infections was well established over the period of time, thus the drug emerged as workhorse anti-MRSA drug.

Vancomycin Intermediate S. Aureus

The antibacterial activity of vancomycin is mediated by its binding to the C-terminal D-Ala-D-Ala residue of the peptidoglycan precursor, and formation of non-covalent complex, thereby, prevents the use of the precursor in bacterial cell wall synthesis. Three decades after its introduction into clinics, no clinical resistance to vancomycin was reported. The first report of a MRSA strain showing reduced susceptibility to vancomycin was reported in 1997. The vancomycin MIC against this strain (Mu50) was 8 mg/L, thus, designated as intermediate sensitive category. The strain had thickened cell wall when observed under electron microscopy and did not carry vanA or vanB genes as found in vancomycin-resistant enterococci (VRE). Subsequently, there were more reports of clinical infections due to MRSA strains with decreased vancomycin susceptibility similar to that of Mu50 strain. The S. aureus strains with a MIC range of 4–8 mg/L are referred to as vancomycin intermediate S. aureus (VISA). There were strains, which showed vancomycin MIC of 2 mg/L but had subpopulation with vancomycin MIC of 4–8 mg/L. These strains are referred to as hetero VISA (hVISA).

The genetic basis of emergence of VISA appears complex. The genetic analysis of VISA strains identified mutations in determinants that control the biosynthesis of bacterial cell wall and/or mutations in the ribosomal gene rpoB. The increased MRSA infection in hospitals has led to extensive use of vancomycin resulting in the selection of MRSA strains with reduced vancomycin susceptibility. The study on prevalence of hVISA and VISA has met with the problem of accurate detection of decreased susceptibility to vancomycin. Different diagnostic methods showed variable sensitivity and specificity leading to contradictory reports in prevalence. During 2010–2014, the prevalence rates of hVISA and VISA among MRSA strain were at 7.01% and 7.93%, respectively. The emergence and increased incidence of hVISA and VISA has limited the therapeutic use of vancomycin in the treatment of MRSA infections in hospital. However, by optimizing the dose regimen and drug delivery, thereby, achieving the desired blood plasma concentration which would give the clinical efficacy is the way forward in preserving the clinical utility of vancomycin.

Vancomycin-resistant S. Aureus

S. aureus strains which are referred to as hVISA and VISA are not considered resistant based on vancomycin susceptibility breakpoint (vancomycin MIC of 8 mg/L) defined by clinical laboratory standards institute (CLSI). Unlike VRE, these strains do not carry vanA or vanB type of genes to confer resistance to vancomycin.

VRSA strains carried copies of the transposon Tn1546, which was acquired from vancomycin-resistant Enterococcus faecalis. The transposon which mediates the VanA-type resistance, encodes a dehydrogenase (VanH), which reduces pyruvate to D-Lac, and the VanA ligase, which catalyzes the formation of an ester bond between D-Ala and D-Lac. The resulting D-Ala-D-Lac depsipeptide replaces the D-Ala-D-Ala dipeptide in peptidoglycan synthesis, a substitution that decreases the affinity of the molecule for vancomycin and other glycopeptide antibiotic, teicoplanin, considerably.

Resistance to other Antibiotics

Since HA-MRSA strains are often MDR phenotype, drugs such as sulphonamides, tetracyclines, aminoglycosides, chloramphenicol and clindamycin were sidelined due to lack of activity, while vancomycin remained the mainstay of therapy. Resistance to sulphonamides and trimethoprim, tetracyclines, aminoglycosides, chloramphenicol and clindamycin, occurring in S. aureus especially among MRSA was widely reported.

Therapeutic Approach

Therapeutic approach to S. aureus infections depends on the type of infection, patient age, clinical manifestation of the disease, co-morbidity, antibacterial susceptibility of infecting organism and hospitalization. Various drugs as single agent and drug combinations have been used to treat S. aureus infection. In general, management of infections due to MRSA is difficult compared to that of MSSA.

Topical Anti-MRSA Drugs

Mupirocin

Mupirocin is used as topical antibiotic to treat impetigo due to S. aureus and S. pyogenes. The drug is also used for nasal decolonization of S. aureus. Mupirocin belongs to monoxycarbolic acid class and it exerts antibacterial action by binding to isoleucyl t-RNA synthetase, thereby, inhibiting the protein synthesis. The antibiotic shows excellent activity against Staphylococci and most Streptococci. Clinical efficacy of mupirocin ointment in treating S. aureus superficial skin infections and wound infections was established. Various reports also demonstrated effectiveness of mupirocin in nasal decolonization of S. aureus that is a risk factor for MRSA infections in nosocomial settings.

Fusidic Acid

Fusidic acid is an antibiotic, which belongs to a class referred to as fusidanes. Chemically it is a tetracyclic triterpenoid and it binds to bacterial elongation factor G (EF-G), which results in impaired translocation process and inhibition of protein synthesis. It has potent activity against S. aureus and clinically used in treatment of mild to moderately severe skin and soft-tissue infections, for example, impetigo, foliculitis, erythrasma, furunculosis, abscesses and infected traumatic wounds. The efficacy of fusidic acid ointment in treatment of S. aureus infections is widely reported. The drug has also been used systemically to treat invasive S. aureus infections but its efficacy was questioned.

Systemic Anti-MRSA Drugs

Vancomycin

Vancomycin remained the mainstay of therapy against MRSA infections in hospitalized patients for decades. Though the antibiotic was available for clinical use since 1958, it gained prominence among clinicians only after the surge in nosocomial MRSA infections in 1980s. Numerous reports documented the clinical efficacy of vancomycin in treating various MRSA infections in hospitalized patients. The emergence and spread of hVISA and VISA strains has threatened the clinical utility of vancomycin. In addition, over the years, the mean MIC of vancomycin against susceptible MRSA populations has increased but within the susceptible range. This phenomenon is referred to as vancomycin MIC creep. There has been poor response to vancomycin therapy in patients infected with vancomycin-susceptible MRSA isolates which had vancomycin MIC at the higher end of susceptible range (2 mg/L). Optimizing the dose regimen and drug delivery, in order to achieve the desired blood plasma concentration which would give the clinical efficacy is the way forward in preserving the clinical utility of vancomycin.

Alternative Therapeutic Approach

Apart from chemotherapeutic approach to tackle the S. aureus infection, alternatives such as agents which inhibit the virulent factors expression and vaccines have been investigated. Various phytochemical are also found to have anti-MRSA activity. All these are at investigational stages and more research is necessary to bring promising candidates for clinical usage.

Antivirulence Agents

Clinical use of agents which are not conventional antibiotics but able to inhibit the expression or function of the virulence factors, rendering the bacteria non-pathogenic is considered an alternative approach to tackle MRSA. Stripping microorganisms of their virulence properties without threatening their existence may offer a reduced selection pressure for drug-resistant mutations. Virulence-specific therapeutics would also avoid the undesirable dramatic alterations of the host microbiota that are associated with current antibiotics.

Accessory gene regulator (agr)-mediated quorum sensing system of S. aureus plays a central role in pathogenesis of Staphylococci. Scientists identified small molecules which inhibited the agr system. Active and passive immunization strategies targeting the virulence factors of S. aureus have also been explored.

Plants

Plants have immune system and other defensive mechanisms against microorganisms that cause plant diseases. Hence, the plants with huge diversity provide a vast source for exploration of anti-MRSA phytochemicals. In vitro Anti-MRSA activity of crude extracts of medicinal plants has been extensively reported. Various phytochemicals such as β-asarone, Mansonone F, prenylated flavonoids and thymoquinone showed in vitro anti-MRSA activity.

Streptococcus Pyogenes

Streptococcus pyogenes is a gram-positive bacterium that usually grows in pairs or chains. It has been classified as a beta-hemolytic streptococcus because when cultured on a blood agar plate all the red blood cells are ruptured by the bacteria. Furthermore, it has been classified using Lancefield serotyping as group A, because it displays antigen A on its cell wall. Therefore, this bacterium is commonly called the beta-hemolytic group A streptococcus, or GAS. A picture of this clearly shows how the organism grows.

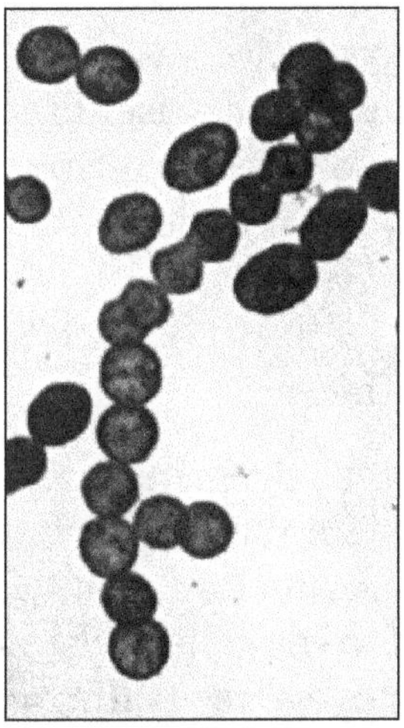

Figure: Electron micrograph of Streptococcus pyogenes.

Streptococcus pyogenes, also known as the flesh eating bacteria, is the most pathogenic bacterium in the whole genus. The name pyogenes comes from the word pyogenic, which is a classification for the streptococci that are associated with pus formation. The effects of this microbe range from mild illnesses such as strep throat and impetigo to more serious diseases such as scarlet fever, glomerulonephritis, and necrotizing fasciitis. However, if strep throat is untreated, it will lead to rheumatic fever, which is pretty rare now in the United States but was a more serious problem before the 20th century.

S. pyogenes usually begins infection on the surface of the skin or in the throat. From there, the bacterium begins to spread into deeper areas of the skin, which can potentially lead to life-threatening diseases. Because of its versatility in the human host, researchers have struggled developing a vaccine that would stop bacterial infection. This is why so much research is being done trying to understand the different cellular components of this organism as well as sequencing the genome. The more characteristics that are discovered and analyzed, the better the understanding we will have to fight this disease. Much progress has been made, because the genome for a strain has already been sequenced, and many conclusions have been drawn from the study.

Genome Structure

The genome of an M1 strain of Streptococcus pyogenes has been sequenced, and was found to contain 1,852,442 base pairs and about 1,752 predicted protein-encoding genes. Also, more than 40 putative virulence associated genes have been identified, and this could explain why Streptococcus pyogenes is considered responsible for a wider variety of human disease than any other bacterial species. GAS was found to produce a large number of extracellular proteins that increase the virulence of the organism by triggering a severe nonspecific immunological response in the human host.

The genome sequence was determined using the whole-genome shotgun approach. It used strain SF370, which was originally isolated from a patient with a wound infection. The genome is a circular chromosome with an average G+C content of 38.5%, and 83% of the genes were transcribed in the clockwise direction and 76% of the genes were transcribed in the counterclockwise direction. However, the replication termination site (ter) has still not been found, but it appears that the site is not 180 degrees from oriC.

Cell Structure and Metabolism

One unique property of Streptococcus pyogenes is that it has a protein called protein F, which is a fibronectin binding protein that allows it to adhere to respiratory epithelial cells. This protein is an important virulence factor because by binding to the epithelial cells, the organism is able to stick to the cells of the host tightly, and not leave. Another characteristic of Streptococcus pyogenes is the M protein, which allows it to resist phagocytosis. The M protein has a fibrillar coiled-coil design, which "offers the organism several distinct advantages, ranging from antigenic variation to multiple functional domains".

In addition, Streptococcus pyogenes is covered with an outer hyaluronic acid capsule. This capsule is required in order for the organism to be resistant to phagocytosis, which is vital in order for it to survive in its hosts. Finally, the "capsule may be an important adherence factor in the pharynx, since it binds CD44 on epithelial cells."

In another study, the regulation of anions such as Pi (inorganic phosphate) was examined in many different microorganisms. It was found that in Streptococcus pyogenes, "32Pi, uptake was catalyzed most rapidly in starved cells and was inhibited by addition of exogenous glycolytic and nonglycolytic energy sources such as glucose and arginine, respectively". This finding was very interesting because the mechanism of the regulation in S. pyogenes is actually the opposite in many other bacteria. The research reports two primary methods of regulation, the first being substrate depletion and the other being cellular ATP. This study is significant because phosphate is very important in regulating the control of many metabolic enzymes. For example, the phosphotransferase system uses a phosphate to transfer glucose into the bacteria by converting it to glucose-6-phosphate.

Pathology

Streptococcus pyogenes is an unusually successful pathogen because of many different properties. First of all, the organism is able to adhere to the cells of its host with strong adhering mechanisms, which is important because the organism would be easily removed by mucus or salivary fluid. The more adhesions there are, the stronger the adherence will be.

Not only does Streptococcus pyogenes adhere to its host cells, but it also invades them. Laboratories have shown that "group A streptococci have the potential to invade human epithelial cells at frequencies equal to or greater than classical intracellular bacterial pathogens." There are two proteins necessary for invasion, which are the M protein and the SfbI, which is a fibronectin-binding protein.

Streptococcus pyogenes affects its hosts in many different ways and causes a large range of diseases. The Center for Disease Control and Prevention (CDC) reports some symptoms of the diseases caused by Streptococcus pyogenes. They range from mild to severe, such as fever, severe pain, dizziness, and red rash at wound site. The symptoms at first might not seem severe, but in some cases the diseases have caused death to certain individuals.

Application to Biotechnology

So far, many of the enzymes released by Streptococcus pyogenes that have been studied are usually involved in how the organism interacts with the host immune system. For example, the cysteine protease SpeB is an enzyme that cleaves the immunoglobulin IgG, which allows S. pyogenes to fight against the host immune system. Another popular immunomodulating enzyme studied is C5a peptidase, ScpA, which is a serine endopeptidase that "cleaves the specific complement factor C5a, inhibits recruitment of phagocytic cells to the infectious site." Since these enzymes all modulate the immune system of the host and are used to help the organism survive, they are most likely not very useful in the field of biotechnology.

However, there has been a patent on two specific proteins that have been found to prevent diseases caused by Streptococcus pyogenes. The two proteins are "designated Cpa1 and Cpa49 which are obtained from the collagen binding region in Streptococcus pyogenes. The isolated proteins of the present invention have been observed to bind to collagen, and thus can be utilized in methods of treating or preventing streptococcal infection through the inhibition of the ability of the bacteria to bind to collagen." This is a perfect example of how a compound extracted from an organism can be used to fight against itself and protect victims from invasion.

Streptococcus Pneumoniae

Streptococcus pneumoniae has been one of the most extensively studied microorganisms since its first isolation in 1881. It was the object of many investigations that led to important scientific discoveries. In 1928, Griffith observed that when heat-killed encapsulated pneumococci and live strains constitutively lacking any capsule were concomitantly injected into mice, the nonencapsulated could be converted into encapsulated pneumococci with the same capsular type as the heatkilled strain. Years later, the nature of this "transforming principle," or carrier of genetic information, was shown to be DNA. Other important discoveries resulting from investigations on pneumococci were the therapeutic efficacy of penicillin, the role of the bacterial capsule in resistance to phagocytosis, the ability of polysaccharides (PS) to induce antibodies (Ab), the first demonstration of antigen-specific tolerance or immunological unresponsiveness, the discovery of regulatory thymus-derived T lymphocytes, and the putative use of PS antigens as vaccines.

Surface Components of S. Pneumoniae

Pneumococci are lancet-shaped gram-positive bacteria, which grow in pairs or short chains. Three major surface layers can be distinguished in their surface: plasma membrane, cell wall, and capsule. The cell wall consists of a triple-layered peptidoglycan backbone that anchors the capsular PS, the cell wall polysaccharide (CWPS), and possibly also proteins. The capsule is the thickest layer, completely concealing the inner structures in exponentially growing pneumococci. Whereas CWPS is common to all pneumococcal serotypes, the chemical structure of the capsular PS is serotype specific.

The capsule consists of high-molecular-weight polymers made up of units of repeating oligosaccharides (OS), which can contain between two and eight monosaccharides. Many serotypes possess acidic components (like D-glucuronic acid or phosphate groups), ribitol, or arabinitol. In six serotypes, phosphorylcholine (PC) is part of the capsular PS. The typical gram-positive cell wall is composed mainly of peptidoglycan: glycan chains of alternating N-acetylglucosamine and N-acetylmuramic acid residues, cross-linked to each other through peptide side chains. The peptide chains have alanine as the first residue, linked to N-acetylmuramic acid. The CWPS, a complex teichoic acid containing PC residues, is also attached to peptidoglycan via N-acetylmuramic acid. The capsular PS is also linked to peptidoglycan. Although the bond between the two components has not yet been identified, it is probably covalent. The PC residues of CWPS are a recognition site for N-acetylmuramic acid-L-alanine amidase, an enzyme that cleaves peptidoglycan into separate glycan and peptide chains by hydrolyzing the bond between alanine and N-acetylmuramic acid. This enzyme is involved in the process of cell division by cleaving the peptidoglycan. This enzyme is also referred to as autolysin, because it induces lysis of pneumococci under certain conditions, i.e., during the stationary phase of growth.

Another important structure is the lipoteichoic acid or Forssman antigen, a teichoic acid similar to CWPS with additional covalently attached lipid material. This antigen, inserted into the plasma membrane via its lipid moiety, also contains PC. The Forssman antigen is a powerful inhibitor of autolysin, and its PC residues are involved in its specific interaction with the enzyme. During the stationary phase of growth, pneumococcal cells release the Forssman antigen. It is thought that the loss of this enzyme-inhibitor, followed by unrestrained activity of autolysin, results in the destruction of the cell wall and finally in bacterial lysis. Immunization of mice with the Forssman antigen does not protect against pneumococcal infection.

Virulence Factors

The pathogenicity of pneumococci has been attributed to various structures, most of which are situated on its surface. The high morbidity and mortality caused by this microorganism are still, however, poorly understood, and the list of virulence factors is probably far from complete. One group of factors, such as the capsule and a recently identified protein, provides resistance to phagocytosis and thus promotes the escape of pneumococci from the host immune defense. Other factors, including cell wall components and the intracellular toxin pneumolysin, are involved mainly in the inflammation caused by infection. The inflammation process probably fully develops only after lysis of bacteria by autolysin. Since inflammation is thought to induce most of the symptoms of pneumococcal disease, this group of virulence factors may thus be more directly responsible for the morbidity and mortality caused by pneumococci once they have infected the host. However, since inflammation can facilitate the further spread of pneumococci to other tissues and organs, this distinction of virulence factors in two groups is obviously not absolute. Recently, the production of hydrogen peroxide by pneumococci and its toxic effects for rat alveolar epithelial cells were demonstrated in vitro. Since the amount of hydrogen peroxide produced by pneumococci is similar to that produced by activated neutrophils, this oxidant might be involved in the pathogenesis of pneumococci in vivo by causing lung injury.

Capsule

The capsule has long been recognized as the major virulence factor of S. pneumoniae. Experimental proof for this was provided by the difference in 50% lethal dose between encapsulated and unencapsulated strains. Encapsulated strains were found to be at least 105 times more virulent than strains lacking the capsule. Recently, pneumococcal mutants that apparently differ only in the type of capsular PS expressed have been produced. The virulence of the mutants in relation to the parental strains was shown to be determined mainly by the capsular type. However, the genetic background of the recipient strain was also of importance, showing that other components besides the capsule are required for full virulence of pneumococci.

The chemical structure of the capsular PS and, to a lesser extent, the thickness of the capsule determines the differential ability of serotypes to survive in the bloodstream and possibly to cause invasive disease. This is probably due to differences between serotypes in terms of activation of the alternative pathway of complement; deposition and degradation of complement components on the capsule, resistance to phagocytosis, ability to induce Ab, and clearance mediated by lectin-like structures on phagocytes.

Table: Characteristics of the major proposed virulence factors of S. pneumoniae.

Virulence factor	Proposed mechanism of virulence
Capsule	Lack of activation of alternative complement pathway. Resistance to phagocytosis. Deposition of opsonically inactive complement components. No or low immunogenicity of some serotypes.
Cell wall or CWPS	Inflammatory effects. Activation of the alternative complement pathway, resulting in anaphylatoxin production. Enhancement of vascular permeability, mast cell degranulation, PMN activation. IL-1 production increased, cytopathic for endothelium. Mediator of attachment to endothelial cells.
Pneumolysin	Cytolytic at high concentrations. Cytotoxic at lower concentrations. Inhibition of ciliary movement and disruption of epithelium. Inhibition of bactericidal activity of PMN. Inhibition of lymphocyte proliferation. Inhibition of Ab synthesis. Complement activation. IL-1b and TNF-a production by monocytes increased, Binding of Fc fragment of Ab.
PspA	Inhibition of complement activation?
Complement factor H-binding component	Inhibition of complement activation. Inhibition of phagocytosis.
Autolysin	Release of pneumolysin and cell wall products.

Neuraminidase	Exposure of "receptors" for pneumococci?
Peptide permeases	Enhancement of adhesion.
Hydrogen peroxide	Lung injury?
IgA1 protease	Counteracts mucosal defense mechanisms?

The mechanisms marked "?" have been suggested but not demonstrated. Most of the mechanisms listed are observed only in vitro. Their significance for pneumococcal virulence remains unknown. IL, interleukin; TNF, tumor necrosis factor. This mechanism of virulence has been demonstrated only for viral neuraminidase, not for pneumococcal neuraminidase.

Hostetter in 1986 proposed an interesting hypothesis that explained some of these differences: capsular serotypes which in some way allow the deposition of the complement component C3b but prevent its degradation to C3d are more easily taken up by phagocytes (via the receptor with high affinity for iC3b, complement receptor 3 [CR3]). As a result, however, they are poorly immunogenic, because they are rapidly cleared. On the other hand, serotypes on which capsule C3b is degraded to both iC3b and C3d are more resistant to phagocytosis but induce a stronger Ab response. Experimental support for the mechanism by which C3d might enhance the anti-PS Ab production, as suggested in 1986 by Hostetter, was provided some years later. It should be noted that this explanation for the phagocytosis and immunogenicity of different serotypes is based on studies with a limited number of serotypes. It is therefore unlikely to apply to all of them. The role of the capsule in the virulence of pneumococci is also well illustrated by the highly protective activity of anticapsular Ab. Recently; survival of mice after challenge with a lethal dose of pneumococci was shown to be predictable from levels of anti-capsular PS Ab present in serum.

Cell Wall and Cell Wall Polysaccharide

In contrast to the capsular PS, purified peptidoglycan and especially CWPS have been found to induce inflammation similar to that seen after infection with whole pneumococci. Typical pneumococcal diseases such as otitis media, meningitis, and pneumonia can be mimicked in animals that have received injection of purified cell wall or its degradation products. CWPS activates the alternative pathway of complement. During complement activation, the anaphylatoxins C3a and C5a, which enhance vascular permeability, induce mast cell degranulation, and recruit and activate polymorphonuclear leukocytes (PMN) at the inflammation site, are produced. Furthermore, purified cell wall is a powerful stimulus (even stronger than endotoxin) for the production of interleukin-1 by human monocytes. This cytokine, together with tumor necrosis factor, plays a pivotal role in the inflammation process. Cell wall also was shown to be involved in the attachment of unencapsulated pneumococci to human endothelial cells and to have (interleukin-1-mediated) cytopathic effects on these cells.

Anti-CWPS or anti-PC Ab has been demonstrated to protect animals against pneumococcal challenge, although such a protective effect was not observed by others. This discrepancy might be due to improper culture conditions of pneumococci, resulting in inocula containing partially unencapsulated bacteria. The protective activity of anti-PC Ab is substantially weaker than that of anti-capsular PS Ab. Furthermore, protection induced by antiCWPS Ab seems not to be mediated by the stimulation of phagocytosis. In humans, no strong differences were observed in levels of anti-CWPS immunoglobulin G (IgG) found in healthy adults and patients at different stages of pneumococcal infection. This suggests that anti-CWPS IgG may not prevent or delay the progress of pneumococcal disease in humans. Nevertheless, since the control individuals in this study were not age matched, the lack of protective effect of anti-CWPS IgG might be due to qualitative differences between IgG from young individuals and IgG from elderly people. Such an age-dependent qualitative difference, in terms of affinity and protective efficacy, has been described for mouse Ab against PC. Whether similar differences play a role in the immune response to pneumococcal PS remains to be established.

Pneumococcal Proteins

Various proteins have been suggested to be involved in the pathogenicity of S. pneumoniae. However, only a few of them have actually been confirmed as virulence factors. Pneumococci produce an IgA1 protease that might interfere with host defense at mucosal surfaces. S. pneumoniae also produces neuraminidase, an enzyme that may facilitate attachment to epithelial cells by cleaving sialic acid from the host glycolipids and gangliosides. Partially purified neuraminidase was observed to induce meningitis-like symptoms in mice; however, the reliability of this finding has been questioned because the neuraminidase preparations used were probably contaminated with cell wall products. Nevertheless, immunization with pure neuraminidase slightly enhances the survival time of mice upon challenge with pneumococci, which confirms the contribution of this protein to the pathogenicity of pneumococci.

Other pneumococcal proteins besides neuraminidase are involved in the adhesion of pneumococci to epithelial and endothelial cells. These pneumococcal proteins have as yet not been identified but are able to recognize glycoconjugates on the surface of host cells. These glycoconjugates contain oligosaccharides with the following specificities: N-acetyl-D-glucosamine-b galactose, N-acetyl-D-galactosamine-b galactose, and glucosamine only. Recently, Cundell et al. reported that peptide permeases can modulate pneumococcal adherence to epithelial and endothelial cells. It was, however, unclear whether these permeases function directly as adhesions or whether they enhance adherence by modulating the expression of pneumococcal adhesions. Other proteins that may enhance the virulence of pneumococci, including hyaluronidase and neutrophil elastase inhibitor. Recently, a protease-sensitive surface component of pneumococci that may be an important virulence factor has been described. This component, likely to be a protein, is able to bind complement factor H, thereby inhibiting complement activation and phagocytosis. Protease treatment of pneumococci enhances their complement-activating ability and complement-dependent phagocytosis and diminishes their virulence in mice. Protease treatment has no effect on the presence of capsule. The production of a C3-degrading protease by exponentially growing pneumococci of different serotypes has been recently suggested by Angel et al. The observed enzymatic activity is heat labile and independent of capsule production; it can be released readily from the pneumococcal surface by mutanolysin treatment.

Pneumolysin

Pneumolysin is an intracellular protein that belongs to the family of thiol-activated toxins. Conserved among pneumococcal isolates, it has a variety of toxic effects on different cell types. Although it is not secreted by pneumococci, it can be released upon lysis of pneumococci under the influence of autolysin. At high concentrations, pneumolysin oligomers are formed on mammalian cell membranes, giving rise to transmembrane pores that cause cell lysis. At lower concentrations, the toxin has several effects, most of which can be demonstrated in vitro. Pneumolysin stimulates the production of inflammatory cytokines like tumor necrosis factor alpha and interleukin-1b by human monocytes, inhibits the beating of cilia on human respiratory epithelial cells, disrupts the monolayers of cultured epithelial cells from the upper respiratory tract and from the alveoli, decreases the bactericidal activity and migration of neutrophils, and inhibits lymphocyte proliferation and Ab synthesis. Furthermore, pneumolysin activates the classical complement pathway in the absence of anti-toxin Ab, an activity that seems to be mediated by binding of Ab via their Fc fragments.

In this regard, it is interesting that pneumolysin has some sequence homology with C-reactive protein (CRP), an acutephase protein that partially protects mice against S. pneumoniae. Upon binding to the PC residues of pneumococcal CWPS, CRP activates the classical complement pathway through binding to C1q. Preliminary data indicate that pneumolysin also may bind C1q directly. Pneumolysin thus may enhance the inflammatory process through its capacity to activate complement, to disrupt the respiratory tract epithelium (thereby facilitating access of pneumococci to the blood), and to stimulate cytokine production. Furthermore, it also may contribute to the virulence of pneumococci by competing with CRP for binding to C1q, thereby abrogating the protective effects of CRP. Pneumolysin-negative mutants are less virulent than their parental strains, and immunization with pneumolysin prolongs the survival of mice after challenge with pneumococci of different serotypes. This toxin may be therefore considered an important virulence factor.

Pneumococcal Surface Protein A

Pneumococcal surface protein A (PspA) is a surface protein with structural and antigenic variability between different pneumococcal strains. It is found in most of the clinical isolates of pneumococci. Since it is probably a transmembrane protein, PspA has proven difficult to purify, and its functions are unknown. Nevertheless, it seems to be required for full virulence of pneumococci. Preliminary experiments in our laboratory indicate that inhibition of complement activation (in a factor H-independent fashion) may be the mechanism by which PspA enhances pneumococcal virulence. Passive immunization with polyclonal or monoclonal Ab against PspA and active immunization with recombinant PspA or its N-terminal fragment have both been shown to protect mice against challenge with pneumococcal strains of various capsular serotypes. These strains probably contain common protective PspA epitopes.

Autolysin

Autolysin-negative mutants have been shown to be less virulent than wild-type pneumococci, and immunization with autolysin confers some protection against pneumococcal challenge in mice. The effects of autolysin, however, seem to be mediated by the release of pneumolysin (and perhaps also cell wall products) from the pneumococcal cytoplasm, because immunization with autolysin does not protect against challenge with a pneumolysin-negative mutant. Furthermore, autolysin was recently shown to be triggered by human lysozyme, a defense factor released upon infection and inflammation, thereby inducing pneumococcal lysis and enhancing the inflammation. Autolysin thus seems to take advantage of the (protective) function of lysozyme.

Different Mechanisms of Virulence

When the mechanisms by which different virulence factors may contribute to pneumococcal pathogenesis are compared, paradoxes are observed. Although CWPS and pneumolysin activate the complement system, the factor H-binding component and PspA inhibit complement activation. This emphasizes our previous suggestion that virulence factors may be divided into two separate groups. One consists of factors present on the surface of intact pneumococci, which seem to act at the beginning of the infection, mainly by impeding phagocytosis via complement inhibition. The second group consists of factors that act at the stage of pneumococcal disintegration and lysis. At this stage, complement activation enhances the inflammation; this appears to be the "point of no return" of pneumococcal infection.

Pathogenesis of Pneumococcal Disease

Carriage and Infection

S. pneumoniae is carried in the upper respiratory tract by many healthy individuals. It has been suggested that attachment of pneumococci is mediated by a disaccharide receptor on fibronectin, present on human pharyngeal epithelial cells. Adherence of pneumococci to tracheal epithelial cells can be enhanced by prior influenza virus infection. This enhancement is thought to be mediated by viral neuraminidase. This enzyme cleaves sialic acid from glycosphingolipids, which is present in substantial amounts in human lung tissue. Thus, neuraminidase may expose other structures that function as receptors for adhering pneumococci.

The mechanisms by which pneumococci translocate from the nasopharynx to the lung, thereby causing pneumonia, or migrate directly to the blood, giving rise to bacteremia or septicemia, are poorly understood. Most infections do not occur after prolonged carriage but follow the acquisition of recently acquired serotypes. This suggests that the immune status of the host at the moment of colonization, as well as the virulence of the particular strain, determines whether pneumococci will remain confined to the nasopharynx or become invasive.

Failure of the specific (secretory IgA) and nonspecific (cough reflexes, mucosal secretion, and ciliary transport) defenses in the respiratory tract may facilitate access of pneumococci to the bronchi and the lungs. The effects of pneumolysin on ciliary beating of epithelial cells and the effects of the IgA1 protease secreted by pneumococci might impair these defense mechanisms. Simultaneously, damage of the epithelial monolayer by hydrogen peroxide (produced by pneumococci) and by pneumolysin may facilitate direct access of pneumococci to the blood. Epithelial damage, caused by previous (viral) upper respiratory infections, also increases the opportunity of pneumococci to reach the bloodstream. From the blood, they may migrate to the meninges and, after disrupting the endothelium, reach the subarachnoid space. Alternatively, they may reach the meninges via other routes, such as directly from the nasopharynx.

Unrestrained multiplication of pneumococci in the lungs, meninges, or middle ear will result in pneumococcal lysis, with the release of cell wall products and pneumolysin. The presence of lysozyme in secretion fluids at the sites of infection may contribute to pneumococcal lysis through the activation of autolysin. Pneumococcal lysis will in turn trigger the inflammatory process, directly by attracting and activating phagocytes and indirectly through complement activation and anaphylatoxin formation. There is increasing support for the hypothesis that such inflammation may be responsible for the morbidity and mortality of pneumococcal infection; this may account for the fact that elimination of pneumococci from the sites of infection by antibiotic therapy frequently does not improve the course or outcome of pneumococcal disease. Moreover, by inducing pneumococcal lysis, highly effective bactericidal antibiotics such as the b-lactams (especially when administered at late stages of infection) may even augment the harmful effects of pneumococcal disease.

Early Response of the Host

Unrestrained proliferation of pneumococci at the site of infection fortunately is not likely to occur in healthy individuals. Alveolar macrophages and infiltrated PMN will clear the bacteria providing specific Ab and complement are present. The importance of opsonization for the removal of pneumococci is strongly supported by the fact that impairment of either the phagocytic system or opsonin production predisposes to pneumococcal infection. Examples of such defects are the lack of Ab (hypogammaglobulinemia, inability to produce anti-PS Ab in children, IgA deficiency, etc., the absence of nonspecific opsonins (complement deficiencies), and defects of the phagocytic system (asplenia, neutropenia, Hodgkin's disease, and others). Once ingested and entrapped in a phago(lyso)some, pneumococci are readily killed, even when PMN lack the ability to produce a normal oxidative burst, e.g., in patients with chronic granulomatous disease. This seems to be due to the inability of pneumococci to resist the toxicity of their own hydrogen peroxide.

Clearance of pneumococci in the absence of specific Ab may be facilitated by CRP-mediated complement activation. However, the anti-pneumococcal effects of CRP are presumably not mediated by the ability to induce phagocytosis, except for serotypes that possess PC in their capsular PS. Moreover, it has been suggested that pneumolysin may counteract the protective effects of CRP. Other non-Ab-mediated mechanisms may involve phagocytosis mediated by lectins, which are proteins that are abundant on macrophages from the liver and spleen and that specifically recognize carbohydrate structures. However, lectinophagocytosis is not likely to be a general mechanism for all pneumococci; rather, it is dependent on the pneumococcal serotype, i.e., the structure of the capsular polysaccharide. Whether phagocytosis is lectin mediated or not, the use of nonimmune guinea pigs as a model has shown that the liver and spleen play a major role in Abindependent pneumococcal sequestration and that an intact complement system is crucial for clearance. The rate of hepatic, as compared with splenic, sequestration of pneumococci was increased both by a functional complement system and by PS-specific Ab. Deposition of complement upon injection of pneumococci in nonimmune animals occurs preferentially on the pneumococcal cell wall because complement is activated by CWPS via the alternative pathway. The finding that, in the absence of anti-capsular Ab, pneumococci is relatively better cleared in the spleen has been explained by a diminished ability of Kupffer cells in the liver to bind C3b deposited on the cell wall; this may be due to the barrier formed by the capsular PS. This explanation, however, has not been supported experimentally.

Antibody-Mediated Clearance

In the presence of anti-capsular Ab, pneumococci are rapidly cleared from the blood, mainly by the liver and to a lesser extent by the spleen; however, complement is necessary to achieve effective clearance. Whether activated by the classical or the alternative pathway, by CRP, Ab, or the PS itself, complement deposition on the capsule (but not on the cell wall) is essential for pneumococcal phagocytosis and clearance. This supports the notion, described above, that the protective effects of anti-CWPS Ab are probably not mediated by phagocytosis of pneumococci. The most likely mechanism by which anti-CWPS Ab confer protection to animals seems to be neutralization of the inflammatory effects of CWPS. A properly functioning complement system and induction of anti-capsular PS Ab able to induce phagocytosis are thus necessary to achieve full protection against pneumococci.

Immune Response to Pneumococcal Antigens

Once an immunocompetent individual has been colonized or infected by S. pneumoniae, an immune response against different bacterial structures will be mounted. These structures include the capsular PS, the CWPS (mainly its PC residues), pneumolysin, and other pneumococcal proteins including PspA. Although Ab against many capsular serotypes has been found in serum from adults, the actual levels of these Ab may have been overestimated because of the codetection of anti-CWPS Ab in many immunoassays. Ab against the capsular PS has the highest protective capacity, with an estimated protective level of 200 to 300 ng of antibody nitrogen (Ab N) per ml of serum in humans. The amount of human anti-capsular PS Ab needed to passively protect mice against 1,000 50% lethal doses of pneumococci was found to be 150 ng of antibody nitrogen (Ab N) per mouse. Ab specific for CWPS, pneumolysin, and PspA also have been shown to confer protection, mostly in animals; however, their protective ability was consistently shown to be less than that of anti-capsular PS Ab. It is also likely that Ab against neuraminidase and autolysin have at least some protective activity.

Detectable levels of IgM against PC are observed in infants after infection with S. pneumoniae, frequently even in the first year of life. Carriage also sometimes results in the development of anti-PC IgM in infants, although the induction of (cross-reactive) anti-PC antibodies by other organisms cannot be excluded. The amount of anti-PC Ab is age dependent: Ab are present in nearly all older children and adults, but their levels decrease after the age of 50 to 60 years. It is likely that infants can elicit anti-capsular PS Ab of the IgG class following pneumococcal infection, although they rarely do so in the case of asymptomatic carriage with poorly immunogenic pediatric serotypes 23F and 19F. Anti-capsular IgG responses have been observed in infants under 2 years of age following H. influenzae type b meningitis. The ability to elicit an anti-PS IgG response against encapsulated bacteria contrasts with the typical responses of infants to purified capsular PS, which are weak and restricted mainly to IgM. Such IgG responses, also observed after immunization of mice and rabbits with killed pneumococci, may be facilitated by association of the capsular PS with proteinaceous surface components of the bacterium. Indeed, pneumococcal capsular PS is covalently linked to peptidoglycan, which may also anchor pneumococcal proteins. The finding that infants develop a clear anti-capsular PS Ab response upon reinfection or reacquisition of the same pneumococcal serotype, even when serotype-specific Ab was present at the time of reexposure, suggests the presence of a type of booster response similar to that seen upon reimmunization with a protein.

Table: Major characteristics of the Ab response to TD, TI-2, and PS-protein antigens.

Characteristic	Effect for:		
	TD antigens	TI-2 antigens	PS-protein conjugates
T cells required	Yes	No	Yes/no[a]
Memory induction	Yes	No	Yes

Response in infants	Yes	No	Yes
Main Ab (sub)isotype[b]			
Humans	IgG1, IgG4	IgM, IgA, IgG2	IgG2, IgG1
Mice	IgG1, IgG2a	IgM, IgG3, IgG1	IgG1, IgG3, IgM

[a] PS-protein conjugates induce TI responses in nude mice, but T cells are required to induce TD responses.

[b] Listed in decreasing order of amount induced.

Thymus-Dependent Antigens

The immune response to protein antigens requires the cooperation of T and B lymphocytes, as shown by the lack of anti-protein Ab formation in nude mice. Therefore, protein antigens are referred to as thymus-dependent or T-cell-dependent (TD) antigens. For the proper activation and differentiation of protein antigen-specific B cells to memory cells or to Ab-producing plasma cells, binding of the antigens to their antigen receptor (membrane Ig) is not sufficient; rather, the interaction between B cells and helper T (Th) cells and cytokine-mediated events are essential. TD responses can be elicited in infants. When healthy infants are colonized or infected by S. pneumoniae, they consequently mount a normal Ab response to pneumococcal proteins, similar to that elicited by adults. TD responses are characterized by the induction of memory as evidenced by a booster effect upon subsequent immunizations, affinity maturation, and extensive Ab (sub) class switching. Primary Ab responses to proteins normally comprise IgM and IgG. The (sub-) classes in secondary Ab responses in humans are, in order of the amount induced, IgG1>IgG4> > IgM> IgG2 ≈ IgA ≈ IgG3. In mice, the isotype distribution of secondary TD responses is IgG1 >IgG2a >>IgG2b ≥IgM >IgG3. IgA levels in serum are almost undetectable.

Thymus-Independent Antigens

In contrast to proteins, most, if not all, pneumococcal capsular PS elicit Ab responses in nude mice and are therefore referred to as thymus-independent (TI) antigens. They are further classified as TI type 1 and type 2 antigens based on their ability to induce Ab responses in CBA/N mice, a strain carrying X-linked immunodeficiency. TI type 2 antigens do not induce an immune response in this mouse strain and are therefore characterized as TI-2 antigens. Typical TI-2 antigens, including various kinds of PS, polypeptides, and polynucleotides, are poorly metabolized polymers with high molecular weight. These antigens are weakly immunogenic in infants younger than 18 to 24 months, and their immunogenicity increases with age.

Although T cells are not required to induce an Ab response against TI-2 antigens, different subsets of T cells are definitely involved in the stimulation or suppression of that response, as has been shown by different approaches. T cells may be activated in different ways to modulate anti-PS responses in an antigen-specific manner. Direct activation of T cells via stimulation of the T-cell receptor by PS, OS, or glycolipids complexed to major histocompatibility complex (MHC) class II molecules is unlikely. Nevertheless, evidence is accumulating that the T-cell receptor can recognize glycopeptides complexed to MHC class II. Recently, the saccharide moieties of glycopeptides were shown to influence the interaction of the peptide with MHC molecules. Another way in which regulatory T cells may be activated is by the recognition of idiotypic determinants from saccharide-specific B cells, as has been demonstrated for anti-dextran IgG responses and has been proposed to occur during the antibody response to pneumococcal PS. T cells, as well as macrophages, mast cells, and natural killer cells, may also influence the response to TI-2 antigens by means of cytokines.

TI-2 responses, in contrast to TD responses, are oligoclonal, dependent on age, and characterized by the lack of affinity maturation, poor Ab (sub) class switching, and the inability to generate memory. Nevertheless, TI-2 antigens are capable of activating preexisting memory cells, as shown by the ability of PS to induce booster responses upon priming with TD PS-protein conjugates. PS induce mainly IgM and IgA, and the IgG subclasses elicited in humans are, in decreasing order, IgG2 >> IgG1≈ IgG3> IgG4. In some studies, IgA (mainly IgA2) was the most important isotype found in serum upon immunization with pneumococcal or other PS. In children, the IgG1/IgG2 ratio induced by PS is inversed: compared with adults, children produce relatively large amounts of IgG1 but small amounts of IgG2. In mice, the anti-PS IgG response was mainly IgG3 >IgG2a. In another study, anti-dextran Ab had the following isotype distribution: IgM >> IgG1 = IgG3 > IgG2a; the isotypes IgG2b and IgA were undetectable.

The molecular mechanisms behind the activation and differentiation of B cells by TI-2 antigens are only partially understood. The activation signals provided by Th cells in the case of TD antigens are supposed to be partially replaced by other factors. PS cross-links membrane Ig on B cells, and C3d, deposited on the PS, is recognized by complement receptor 2 (CD21, CR2) on B cells in vitro. These two signals increase the numbers of B cells secreting anti-PS IgM. Although the C3d-mediated activation mechanism is not definitely demonstrated to occur in vivo, its potential significance is supported by the finding that B cells of the marginal zone of the spleen in neonates have a low expression of CR2. Marginal-zone B cells are involved in the initiation of TI-2 responses, which, in contrast to TD responses, do not frequently occur in infants younger than 2 years of age. The delay in the ontogeny of the immune response against these antigens may thus be explained by the inability of neonatal B cells to be triggered via CR2, a signal that is necessary for full activation. Interestingly, CBA/N mice, which do not respond to TI-2 antigens and therefore resemble neonates, also have a low CR2 expression on B cells.

Gram-negative Bacteria

These are bacteria that define the opposite of the gram-positive bacteria in relation to the differential staining technique. During gram staining, the gram-negative bacteria will lose the crystal violet dye color after a wash with alcohol and take up the pink/red color of the counterstain, safranin.

The two classes of bacteria are differentiated through gram staining, because of their cell wall composition i.e Gram-positive bacteria have a large layer of peptidoglycan and a thin layer of the lipid layer, and unlike the Gram-negative bacteria which have a thick layer of lipids and they lack the peptidoglycan layer or some have a very thin layer of the peptidoglycan layer. The absence of peptidoglycan makes their cell-wall less strong, and hence the primary stain is easily washed off by alcohol and water.

Gram-negative bacteria are found in a wide range of environments especially those that have life, ranging from normal flora such as Escherichia coli to pathogenic bacteria such as Klebsiella pneumoniae, Chlamydia trichomatis.

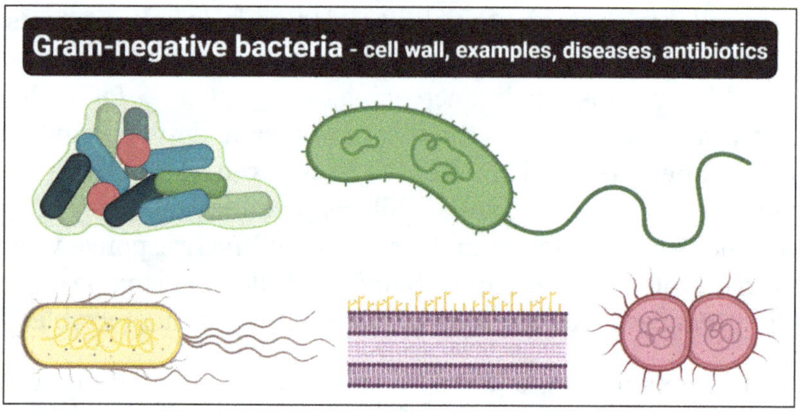

Characteristics of Gram-negative Bacteria

The gram-negative bacteria have a characteristic cell envelope that differentiates it from the Gram-positive bacteria. They have a periplasmic space (Periplasm) which separated the outer membrane from the cytoplasmic layer. The periplasmic space has a network of peptidoglycan chains known as the peptidoglycan layer. They have a thin peptidoglycan layer. They have an outer membrane that is made up of lipopolysaccharides.

Characterization of Gram-negative Bacteria Shapes

Microscopic observation of Gram-negative bacteria cells ranges from rods to bacillus, to cocci to spiral shapes which are the most commonly known shapes. However some exhibit special shapes such as coccobacillus, tetrads, palisades, trichomes, etc.

Rob/bacillus shaped – Escherichia coli. Coccobacillus – which is a combination of both cocci and bacilli shapes include Hemophilus influenza. Streptobacilli are rod-shapes that are connected together in chains eg Streptobacillus moniliformis. Trichome shape is a series of rod-shaped cells that are arranged in a columnar form and it may be enclosed in a sheath. Spiral shaped bacteria also called spirochaetes e.g Chlamydia trachomatis, Treponema pallidum. Filamentous shaped gram-negative have a filament-like shape eg Norcadia spp.

Gram-negative Bacteria Cell Wall

The Cell wall of the Gram-Negative Bacteria is very complex as compared to that of Gram-Positive Bacteria. Combined with the major role of the outer membrane of the cell, with a layer of peptidoglycan, its functional properties are complex, and here is a description of the cell wall and its functional parts.

The cell wall of gram-negative bacteria is complex having a thin layer of the peptidoglycan layer of 2-7nm and a thick outer membrane of 7-8nm thick. Microscopically, there is a space that is seen between the cell membrane and the cell wall, known as the periplasmic space made up of periplasm. However it is found in both Gram-negative and Gram-positive bacteria, but in gram-negative the periplasmic space is larger.

The Periplasmic Space

The periplasmic space in Gram-negative bacteria is made up of several proteins that assist in acquiring nutrients such as hydrolytic enzymes that attack the nucleic acids and phosphorylated molecules, and the binding proteins which actively assist in the transportation of materials into the bacteria cell. The periplasmic space also has enzymes that synthesize peptidoglycan and modify toxic elements that may cause harm to the cell.

Peptidoglycan

The Gram-negative bacterial cell wall has a thin peptidoglycan layer that is found above the plasma membrane making about 5-% of the cells dry weight. Some bacteria like E. coli have 2nm thick peptidoglycan (2-3 sheets of peptidoglycan).

The Outer Membrane and the Lipopolysaccharide

The outer membrane lies above the thin peptidoglycan layer. The membrane is made up of Braun's lipoprotein which is covalently bound to the peptidoglycan and embedded to the outer membrane by hydrophobic ends. The Braun's lipoprotein strongly links the outer membrane and the peptidoglycan. The gram-negative cell wall is also strengthened by the adhesion sites on the outer membrane which play a role in allowing cell contact and membrane fusion. Substances enter the cell through these adhesion sites.

The outer membrane is majorly made up of Lipopolysaccacharides (LPSs) which are large complex molecules made up of lipids and carbohydrates. The lipopolysaccharide is made up of 3 units: Lipid A, the core polysaccharides and the O side chain. The most researched gram-negative bacterial lipopolysaccharide is that of Salmonella Typhimurium. The lipid A unit constitutes two glucosamine sugar derivates each containing three fatty acids and pyrophosphate, Any left parts of the lipopolysaccharide project out of the surface of the membrane.

The O side chain is also known as the O antigen is a chain that extends from the core outwardly. It is made up of sugars that cause variations between bacterial strains. These O antigens are also responsible for bacterial evading antibody responses. The Lipopolysaccahrides are also responsible for protecting the cell wall from external attacks. The LPSs have a negative charge giving the cell surface a negative charge. Thus this stabilizes the membrane structure. Lipid A contributes to the toxic element of the Lipopolysacchaides, hence it acts endotoxins. The outer membrane of the gram-negative bacteria and its lipopolysaccharide contribute greatly to preventing the entry of antibiotics, bile salts and other toxic elements from entering and disrupting the cell. Consequently, the outer membrane is made up of porin proteins which makes it permeable allowing the entry of small molecules such as glucose. Larger molecules like Vitamin B12 are transported across the outer membrane by specific carriers. The outer membrane also contributes to the prevention of loss of components, especially from the periplasmic space.

Gram-negative Bacteria List and their Pathologies and Clinical Significance

Gram-negative bacteria are known to be normal flora apart from some which cause severe human infections which range from community-acquired infections to nosocomial (Hospital-acquired) Infections. They cause severe infections some of which cause death if not detected and treated on time. Here are examples of some of the Gram-negative bacteria and the clinical features they exhibit upon causing diseases and infections in the human host.

Gram-negative bacteria	Pathologies: Clinical feature
Neisseria gonorrhoea	Genitourinary tract infections for both male (urethra) and female (vagina and endocervix), associated with purulent urethral discharge and painful urination. The infection may progress to the uterus, causing salpingitis (inflammation of the fallopian tubes), pelvic inflammatory disease (PID), and fibrosis. Infertility may occur for women with salpingitis. Renal infection for men associated with constipation, painful defecation and purulent discharge. Pharyngitis if purulent pharyngeal exudation occurs. Ophthalmia neonatorum in newborns acquired during birth through the cervix. Disseminated infection associated with fever; painful, purulent arthritis; and small, single, scattered pustules on the skin, with an erythematous (red) base due to dilation or congestion of capillaries. Necrosis may develop.

Neisseria meningitidis	Meningitis can rapidly spread causing meningococcemia if the bacteria invades the bloodstream, associated with a high fever. It can also spread and invade the brain barrier causing purulent meningitis with fever and severe headaches, joint aches and a petechial and/or purpuric rash. Septicemia occurs within the first 12 hours of infection which can progress to fulminant septicemia and shock especially in children known as Waterhouse-Friderichsen syndrome.
Escherichia coli	Intestinal diseases include; enterotoxigenic (ETEC), enteropathogenic (EPEC), enterohemorrhagic (EHEC), enteroinvasive (EIEC), and enteroaggregative (EAEC). They are all associate with diarrhea (watery or/and bloody). Extraintestinal diseases such as Urinary tract infection including cystitis and pyelonephritis, neonatal meningitis and nosocomial-acquired infections such as sepsis/bacteremia, endotoxic shock, and pneumonia.
Salmonella spp	Enteric and Typhoid fever characterized by fever, abdominal pain, and severe symptoms include chills, sweats, headache, anorexia, weakness, sore throat, cough, myalgia, and either diarrhea or constipation. Gastroenteritis (salmonellosis) characterized by nausea, vomiting, and non-bloody diarrhea. Bacteremia associated with abdominal infections (of the hepatobiliary tract and spleen); osteomyelitis; septic arthritis.
Campylobacter jejuni	It causes intestinal and extraintestinal disease with systemic associated with fever, headache, myalgia, and abdominal cramping and diarrhea, which may or may not be bloody. It commonly causes traveler's diarrhea and pseudoappendicitis with symptoms simulating appendicitis without inflammation of the appendix. Bacteremia (often transient) may occur, most often in infants and older adults.
Salmonella dysenterae	Shigellosis (bacillary dysentery) characterized by diarrhea with blood, mucus in stool, and painful abdominal cramping.
Vibrio cholerae	Cholera associated with profuse watery diarrhea massive loss of fluid and electrolytes from the body.
Helicobacter pylori	Acute gastritis with diarrhea. Superficial gastritis associated with epigastric discomfort causing both duodenal ulcers and gastric ulcers. Persistent ulceration may lead to mucosa associated-lymphoid tumors.
Klebsiella pneumoniae	It causes UTI and nosocomial acquired bacteremia.
Pseudomonas aeruginosa	It causes opportunistic nosocomial infections of wounded patients from surgeries, invading the body through catheters and respirators. Keratitis and endophthalmitis after an injury that formed a wound, associated with necrotic otitis. Skin wound infections. Respiratory tract infections characterized by pneumonia symptoms. Gastrointestinal infections with diarrhea. Necrotic enterocolitis in infants. Systemic infections associated with septicemia, pneumonia, bone and joint infections, Infection of the central nervous system (CNS) and soft tissue infections in hospitalized patients.

Gram-negative Bacteria Antimicrobial Agents

Antimicrobial agents against Bacteria causative agents are known as antibiotics. These antibiotics against the mechanisms of the bacterial cell blocking or inhibiting certain mechanisms from being initiated to induce bacterial cell multiplication and replication. Examples of antibiotics used against gram-negative bacteria are listed in the table below:

Antibiotic	Antibiotic	Bacterial agent
Cephalosporin: ceftriaxone	Disruption of the cell by binding to the penicillin-binding proteins and enzymes responsible for the synthesis of peptidoglycan.	Neisseria gonorrhea Neisseria meningitides Pseudomonas aeruginosa
Tetracycline: Doxycycline	Inhibit protein synthesis by preventing the elongation of polypeptides at 30s ribosomes.	Neisseria gonorrheae
Streptogramins	Inhibition of protein synthesis by preventing polypeptide elongation of 50s ribosomes.	Neisseria gonorrheae
B -lactam: Penicillin G	Inhibition of cell synthesis by disrupting the penicillin-binding proteins and enzymes used for peptidoglycan synthesis.	Neisseria meningitidis Pseudomonas aureginosa
Rifampin: Rifamycin	Inhibition of nucleic acid synthesis by preventing transcription of binding DNA-dependent RNA polymerase.	Neisseria meningitides Escherichia coli
Macrolide: Erythromycin, Azithromycin, Clarithromycin	Inhibition of bacterial protein synthesis by preventing the elongation of polypeptides of 50s ribosomes.	Neisseria gonorrhea Campylobacter jejuni Shigella dysenterae Helicobacter pylori Pseudomonas aeruginosa
Tetracycline: Doxycycline	Inhibition of protein synthesis by preventing the polypeptide elongation of 30s ribosomes.	Neisseria gonorrheae Vibrio cholerae
Quinolones: Fluoroquinolones, Ciprofloxacin	Inhibition of nucleic acid synthesis by binding to the alpha-subunit of DNA gyrase.	Eschericia coli Salmonella typhi/paratyphi (enteric and typhoid fever) Campylobacter jujeni Shigella dysenterae Pseudomonas aureginosa
Aminoglycosides	They inhibit protein synthesis by prematurely producing aberrant peptide chains from 30s ribosomes.	Escherichia coli (localized and systemic infections)

Sulfonamides: sulfamethoxazole	They are antimetabolites that inhibit dihydropteroate synthase and distract folic acid synthesis.	Eschericia coli (UTIs and systemic diseases)
Trimethoprim	They are antimetabolites that inhibit the action of dihydrofolate reductase and disrupts the folic acid synthesis.	Escherichia coli(UTIs)

Escherichia Coli

Escherichia coli is the only member of the genus Escherichia and its biochemical properties are now well defined. Subdivision of the species for epidemiological purposes or to assess pathogenicity is performed by means of a serological typing system involving O, K and H antigens. This typing system is complex; a large number of antigens are involved and cross-reactions between antisera and heterologous organisms regularly occur. For these reasons full E. coli serotyping is performed in only a small number of laboratories. The typing system, although laborious and time consuming, gives a high level of discrimination and shows good reproducibility. Typability is good with European strains from human infection but falls with strains from other sources. Recently a fimbrial typing system has also been introduced. Because the serological typing scheme for E. coli is so laborious other typing methods have been attempted. These include phage, colicin and chemo-typing. None have been widely used.

Serotyping is most commonly performed in a attempt to assess pathogenicity in infants with diarrhoeal disease, although it may also be used to assess pathogenicity in other sites, to distinguish between re-infection and recrudescence of infection, particularly in the urinary tract, and for epidemiological studies.

E. coli may also be characterized by a number of properties including production of enterotoxins and haemolysin and by ability to adhere and invade. Such characteristics have been particularly studied in relationship to the pathogenesis of infection in man and animals.

Distribution and Methods of Spread

Environment

The distribution of E. coli in the environment is determined by its presence in the bowel of man and animals, whilst its isolation from water supplies is used as an indicator of recent faecal pollution and of the likelihood of contamination by bowel pathogens. The methods for the detection of E. coli are well-established and standardized as are those for the differentiation of E. coli from other coliforms which may be present, but which are of less significance.

Faecal pollution of natural waters used for recreational purposes has become of increased importance and the consequences of the discharge of untreated sewage into river estuaries and the sea have been extensively studied. However local conditions at different times considerably affect the results, and there is not a great deal of evidence of disease arising from recreation and water sports except where there is gross fouling. Less direct routes by which pathogens may be transmitted from contaminated water include the eating of contaminated shellfish and use of such water in vegetable growing. The situation in soil is very similar to that with water in that the density of E. coli present is directly related to levels of faecal contamination.

Colonization of Human Infants and Young Animals

Under natural conditions many animals are very rapidly colonized by E. coli, large numbers being present in the caecum after the end of the first day of life. However they are not essential to the normal development of the gut and laboratory animals kept under defined conditions may appear normal in the absence of a coliform population. Some normal human beings have no readily detectable E. coli in the faeces. Human infants acquire their coliforms predominantly from the mother but in maternity units some strains spread and colonize a number of babies.

Colonization of Adults

The human faecal coliform flora is heterogeneous and there may be variations with time and also with site in the intestine. The presence of dominant strains in the bowel and there has been a widely held view that such dominant strains may persist for long periods. The persistance of strains in the bowel is however also related to the frequency and numbers of E. coli ingested in food. The organism is commonly found in food prepared in hospitals, canteens and schools and is also frequent in retail processed food. In all these situations the numbers present may be sufficient to colonize the bowel and it appears probable that the serotypes of E. coli present in the human bowel reflect not only the ability of individual strains to colonize but also the coliform population of the diet.

A number of properties may be important in the ability of E. coli strains to establish in the bowel. The possibility that multiply antibiotic-resistant strains may be less able to establish than sensitive strains requires further investigation. There is considerable evidence that colicin production and sensitivity are factors but that they are not of overwhelming significance. Competition for available nutrients, fatty acid production and differences in growth rate may all play a part. Adhesion is likely to be of importance particularly when there is a rapid transit time.

Some serotypes appear to be more likely to persist than others and have been regarded as 'strong'. Whether this is due to the serotype or to a related property is not known. An important aspect to be considered is how well animal strains are able to implant in the human bowel as this has implications as regards the importance to man of antibiotic usage in animals.

Serotype Distribution

There is a shortage of information about E. coli serotype distribution. Evidence exists for geographical variation in the normal human faecal flora even over quite short distances and also for variation in the serotypes that act as pathogens. Although some serotypes are found in both the animal and human populations the overall distribution of types is different. There is little information about serotype distribution in animals and this might be valuable if it could be related to that of the human population.

Antibiotic usage in Animals and its Consequences

There has been a great deal of discussion about the importance to man of the use of antibiotics in animals and, although there is insufficient evidence to come to a conclusion about the significance of antibiotic-resistant coliforms reaching the human population, it is perhaps worth considering some of the factors involved.

There is no doubt that the coliform population of animals may be highly antibiotic-resistant. Many animal pathogens are known to reach man and interest has centred mainly on the salmonellas which have caused major outbreaks in the human population. However, in developed countries normally commensal organisms such as E. coli are responsible for the majority of infections. There is good evidence that animal strains of E. coli do reach the human population by the same route as salmonellas. That is, they are brought into the kitchen on raw meat and are either not destroyed by cooking or cross-contaminate to cooked food. Coliforms may also be transferred between domestic animals and their handlers. In the case of pets it is known that some dogs may have an antibiotic-resistant coliform flora but there is little evidence that this is important to man. Although there is some evidence that animal strains may survive less well in the human gut than human strains, many will persist for several days and some for many weeks after ingestion.

If the survival of antibiotic-resistant organisms following ingestion by man is considered there is again a shortage of information. Much of the work has involved short term studies and we know little about the long term persistance of antibiotic-resistant coliforms in the bowel of man particularly following cessation of antibiotic administration. Antibiotic-resistant animal coliforms which reach the human population may be directly important as pathogens themselves or they may transfer antibiotic resistance to other pathogenic bacteria. The most common human disease caused by E. coli is urinary tract infection and a consideration of the likelihood of animal strains causing such infection in man involves not only the likelihood of their reaching the urethra and bladder but also whether they possess the properties which are associated with the ability to cause urinary tract infection. There is little information on this point. Transfer of antibiotic resistance may occur as was demonstrated by Williams Smith but the transfer was small and it is known that the large number of bacteroides present in the gut does mechanically inhibit transfer. We have therefore no real means of assessing the importance of animal strains of E. coli for man and unfortunately there appear to be no simple ways of making this assessment.

Escherichia Coli and Human Disease

Urinary Tract Infection

In domiciliary practice urinary tract infections are commonly seen in young women. E. coli is frequently isolated and the condition generally responds rapidly to antibiotic therapy, although it is liable to recur and may be distressing and disabling. Urinary tract infection in otherwise healthy young women is unlikely to have important sequelae but in pregnancy bacteriuria is not only more common but may be associated with pyelonephritis, prematurity and foetal loss, although the results obtained by different workers studying this condition have varied.

In the newborn the symptoms of urinary tract infection are apathy, failure to feed and slow weight gain and such infections, and indeed all urinary tract infections occurring before school-age, are liable to result in chronic renal damage. Urinary tract infections in hospitals generally follow catheterization although spontaneous infections also occur. In domiciliary practice E. coli is responsible for the vast majority of urinary tract infections whereas in hospitals, particularly following surgery, other organisms may be involved.

Although there is no doubt that host factors are ofimportance in the development of urinary tract infection it appears highly probable that the characteristics of the invading organism are also of significance. Some workers have, however, considered that the strains causing urinary tract infection merely reflect those commonly present in the bowel. As the bowel is generally recognized to be the source of strains causing urinary tract infection, comparison of faecal and urinary isolates has been used to demonstrate whether selection has occurred.

Comparison of a heterogeneous faecal coliform population with homogeneous urinary isolates adds to the complexity. The characteristics which have been considered may be of significance include serotype, K antigen content, motility, fimbriation, adhesion, invasion, haemolysin and colicin production, serum sensitivity and dulcitol fermentation.

The view that certain 'O' types are particularly associated with urinary tract infection has had support for many years. Generally, although the o types causing infection are also regularly found in faeces, a small number of types cause most infections. These o types include o2, o4, o6, o18 and o 75. Whether these o types are more liable to cause infection due to some property directly associated with the o antigen itself or whether it is because other properties are linked to that o type is not clear. This is a general problem in considering the relationship of E. coli serotype to pathogenicity.

K antigens may be involved in the development of urinary tract infection in a number of ways. Capsulate strains have been found to be more common in urinary than in faecal isolates and in addition strains associated with disease of the upper urinary tract may produce larger amounts of K antigen than strains from infections limited to the bladder. As well as the amount of K antigen the type may also be important. K 1 and K2 are common amongst pyelonephritic strains.

Haemolysin production is one ofthe properties first associated with pathogenicity in the urinary tract. A number of workers have found urinary strains to be more commonly haemolytic than faecal strains; the haemolysins enable the organism to obtain iron from red blood cells, although they may also be associated with a cytotoxic action seen in tissue cultures.

E. coli which gain access to the urinary tract are liable to be washed out and fail to establish themselves unless they are able to adhere to the urinary epithelium and much interest has recently been focussed on the ability of E. coli in the urinary tract and bowel to adhere to host cells. Adhesion involves an interaction between the invading organism and host receptors and study of this phenomenon may throw light not only on bacterial virulence factors but also on host susceptibility to infection.

Cell surface structures that mediate adhesion are termed adhesins and the best studied of these are the fimbrial adhesins. Their presence is often demonstrated by red cell agglutination which may be sensitive to the presence of small amounts of mannose (MSHA) or resistant to it (MRHA).

MRHA is important in urinary tract infection as many strains with this property bind to the widely distributed P blood group antigen. P fimbriate E. coli are particularly found in infections of the upper urinary tract. Other MRHA bind to different receptors but the importance of these adhesins in urinary tract infection has not been fully assessed. Strains with MSHA bind to mucus and their role in urinary tract infection is ill-understood. Their elimination may, in fact, be assisted by mucus binding. Adhesion in the urinary tract has also been studied by examining the ability of E. coli to bind to cells including urinary epithelial cells. Urinary E. coli adhere more commonly than faecal strains. The relationship between adhesion and haemagglutination is complex depending on the presence of similar receptors in the erythrocytes and epithelial cells. Following adhesion, invasion of cells by some urinary E. coli has been reported by one group of workers. The development of knowledge of the adhesive properties of E. coli and particularly the ability to define the P receptor has advanced understanding of bacterial pathogenicity and has led to a great upsurge of interest in bacterial adherence in a number of bacterial species.

Serum sensitivity studies show no difference between faecal and urinary strains nor does colicin production appear to be important, but there have been reports that dulcitol fermentation may be of significance perhaps because the gene is situated near that of a virulence determinant and their expression linked.

The association of serotype and enteropathogenicity and also enterotoxin production and invasiveness is well established, as is that of serotype and antibiotic resistance. In the case of urinary strains the association between the serotypes common in urinary tract infection and haemolysin production is also well known. There have been, in addition, some reports of the association of these serotypes with the K1 antigen, high K antigen production and adhesion to urinary epithelial cells in strains from this source. The relationship of these properties to the ability to cause urinary tract infection is not clear cut. No single property is consistantly found and, although the properties of the group may vary, individual strains may be indistinguishable from faecal strains. This possibly reflects the importance of host factors. Certain individuals may be so predisposed to infection of the urinary tract that strains with few virulence factors may be able to establish themselves. In individuals less predisposed to infection it may be that rather than a single attribute some association of properties is required, strains able to cause infection possessing differing numbers and types of uropathic attributes.

Diseases of the Gastrointestinal Tract

It is in the study of E. coli gastro-enteritis, particularly in infants but also in adults that major advances have been made in our understanding of bacterial pathogenicity in the gut. But there are some other diseases of the gut in which a study of the coliform flora is also of interest.

The aetiology of acute appendicitis is probably multi-factorial and it appears unlikely that bacteria play more than a contributory role. There have been a number of reports that the E. coli flora of the inflamed appendix differs from that of the normal bowel and it has been suggested that the strains present resemble those from urinary tract infection. They are commonly haemolytic, belong to a small number of serotypes, adhere to HeLa cells and are serum resistant. Whether these E. coli strains have any involvement in the pathogenesis of the condition, or whether the differences occur because conditions in the inflamed appendix predispose to the selective overgrowth of previously minority populations with these properties, is not known.

The same problem occurs in relation to ulcerative colitis. This chronic, relapsing inflammatory disease of the large bowel is associated with E. coli strains that are haemolytic, belong to a small number of serotypes and adhere to and invade HeLa cells. Such evidence as there is suggests that these changes may follow rather than initiate relapse of the condition. It is also of interest that, whereas the normal faecal coliform flora is highly heterogeneous, the flora in the acutely inflamed appendix and in the diseased colon shows a greater degree of homogeneity.

There have also been reports that E. coli with a variety of enteropathogenic properties may be associated with the first attack of the ulcerative colitis before the condition is fully developed clinically.

Wound Infection

E. coli is a common cause of wound infection particularly following operations on the bowel when anaerobic organisms may also be present. The most frequent source of the infecting organism is the patient's own bowel. Strains from wound infections resemble in some ways those from the urinary tract in that they tend to belong to a small number of serotypes, produce haemolysin and are serum resistant. Again the coliform flora is homogeneous and it appears that selection from the heterogeneous faecal flora has probably occurred.

Studies of the possibility that particular strains of E. coli may cross-infect post-operative wounds have not so far been reported although there have been suggestions that a nosocomial E. coli flora exists. The absence of information of this sort is probably related to the complexity of the typing system.

Septicaemia

It might be expected that differences between infecting strains of E. coli and those present in the faeces might best be demonstrated by studying strains from septicaemia and meningitis, conditions in which the clinical significance of the isolates cannot be doubted. E. coli is a common cause of septicaemia. Host factors are of outstanding importance but strains from septicaemia again demonstrate properties which clearly distinguish them from faecal strains. They commonly belong to a small number of serotypes, are K antigen-rich, serum resistant, haemolysin and colicin producers and ferment dulcitol. Shock and death following septicaemia is associated with certain serotypes (o4, o6 and o8) and with serum resistance.

The source of the invading organism is often the urinary tract so that some of the properties may be related to the source of the strain. However, there are differences notably, as would be expected, in serum resistance but also in colicin production and dulcitol fermentation.

Meningitis

Coliform meningitis is particularly a disease of neonates being responsible for about a third of cases in this age group. Differences between meningitic and faecal strains are again clear. The prevalent o groups are o1, o6, o7, o16, o18 and o 83 and all these are associated with the K 1 antigen when isolated from meningitis. The relationship between K 1 and the disease is particularly clear, 80 % of the strains having this antigen. The neonatal faecal carriage rate is much lower.

In adults meningitis is less common and is associated with serotypes o 4 and o 75 but the K 1 antigen, which has the same structure as Neisseria meningitis B capsule, is not significant. Capsules not only enable the organism to evade non-specific host defences but may also evade recognition by the host, perhaps because of similarities with host cell-surface structures. K 1 is an antigen which is able to do this.

Pneumonia

Although coliform pneumonia appears to have become of increasing importance during recent years this is a condition on which the assessment of the significance 529 of the organism isolated is very difficult unless samples are taken directly by transtracheal aspiration. There appears to be little data on the properties of E. coli associated with pneumonia.

E. Coli and Disease in Animals

In addition to causing disease in farm animals, E. coli is also responsible for illness in pets and in wild animals although these are of much less economic importance and are less well studied than in farm animals. In puppies enteric and systemic colibacillosis may occur and E. coli also causes urinary tract infection in dogs. It is interesting that although dogs have lived in close contact with man for thousands of years their coliform flora is distinct. Strains causing urinary infection in dogs, as in man, are commonly haemolytic suggesting that this may be an important characteristic.

Salmonella Spp

Salmonellae are ubiquitous human and animal pathogens, and salmonellosis, a disease that affects an estimated 2 million Americans each year, is common throughout the world. Salmonellosis in humans usually takes the form of a self-limiting food poisoning (gastroenteritis), but occasionally manifests as a serious systemic infection (enteric fever) which requires prompt antibiotic treatment. In addition, salmonellosis causes substantial losses of livestock.

Some infectious disease texts recognize three clinical forms of salmonellosis: gastroenteritis, septicemia, and enteric fevers. The septicemic form of salmonella infection can be an intermediate stage of infection in which the patient is not experiencing intestinal symptoms and the bacteria cannot be isolated from fecal specimens. The severity of the infection and whether it remains localized in the intestine or disseminates to the bloodstream may depend on the resistance of the patient and the virulence of the Salmonella isolate.

The incubation period for Salmonella gastroenteritis (food poisoning) depends on the dose of bacteria. Symptoms usually begin 6 to 48 hours after ingestion of contaminated food or water and usually take the form of nausea, vomiting, diarrhea, and abdominal pain. Myalgia and headache are common; however, the cardinal manifestation is diarrhea. Fever (38 °C to 39 °C) and chills are also common. At least two-thirds of patients complain of abdominal cramps. The duration of fever and diarrhea varies, but is usually 2 to 7 days.

Enteric fevers are severe systemic forms of salmonellosis. The best studied enteric fever is typhoid fever, the form caused by S typhi, but any species of Salmonella may cause this type of disease. The symptoms begin after an incubation period of 10 to 14 days. Enteric fevers may be preceded by gastroenteritis, which usually resolves before the onset of systemic disease. The symptoms of enteric fevers are nonspecific and include fever, anorexia, headache, myalgias, and constipation. Enteric fevers are severe infections and may be fatal if antibiotics are not promptly administered.

Structure, Classification and Antigenic Types

Salmonellae are Gram-negative, flagellated, facultatively anaerobic bacilli possessing three major antigens: H or flagellar antigen; O or somatic antigen; and Vi antigen (possessed by only a few serovars). H antigen may occur in either or both of two forms, called phase 1 and phase 2. The organisms tend to change from one phase to the other. O antigens occur on the surface of the outer membrane and are determined by specific sugar sequences on the cell surface. Vi antigen is a superficial antigen overlying the O antigen; it is present in a few serovars, the most important being S typhi.

Antigenic analysis of salmonellae by using specific antisera offers clinical and epidemiological advantages. Determination of antigenic structure permits one to identify the organisms clinically and assign them to one of nine serogroups (A-I), each containing many serovars. H antigen also provides a useful epidemiologic tool with which to determine the source of infection and its mode of spread.

Table: Ecologic Classification of Salmonellae.

Species	Representative Serovar(s)	Reservoir (Host preferences)
S choleraesus	One only	Animals(swine)
S typh	One only	Humans
Maple Academy	Paratyphi-A	Humans
	Schottmuelleri	
	Pufforum	Animals(fowl)
	Dublin	Animals(cattle)
	Typhimurium	
	Derby	
	Enteritidis	Humans and many animals
	Heidelberg and hundreds of related serovars	

As with other Gram-negative bacilli, the cell envelope of salmonellae contains a complex lipopolysaccharide (LPS) structure that is liberated on lysis of the cell and, to some extent, during culture. The lipopolysaccharide moiety may function as an endotoxin, and may be important in determining virulence of the organisms. This macromolecular endotoxin complex consists of three components, an outer O-polysaccharide coat, a middle portion (the R core), and an inner lipid A coat. Lipopolysaccharide structure is important for several reasons. First, the nature of the repeating sugar units in the outer O-polysaccharide chains is responsible for O antigen specificity; it may also help determine the virulence of the organism. Salmonellae lacking the complete sequence of O-sugar repeat units are called rough because of the rough appearance of the colonies; they are usually avirulent or less virulent than the smooth strains which possess a full complement of O-sugar repeat units. Second, antibodies directed against the R core (common enterobacterial antigen) may protect against infection by a wide variety of Gram-negative bacteria sharing a common core structure or may moderate their lethal effects. Third, the endotoxin component of the cell wall may play an important role in the pathogenesis of many clinical manifestations of Gram-negative infections. Endotoxins evoke fever, activate the serum complement, kinin, and clotting systems, depress myocardial function, and alter lymphocyte function. Circulating endotoxin may be responsible in part for many of the manifestations of septic shock that can occur in systemic infections.

Pathogenesis

Salmonellosis includes several syndromes (gastroenteritis, enteric fevers, septicemia, focal infections, and an asymptomatic carrier state). Particular serovars show a strong propensity to produce a particular syndrome (S typhi, S paratyphi-A, and S schottmuelleri produce enteric fever; S choleraesuis produces septicemia or focal infections; S typhimurium and S enteritidis produce gastroenteritis); however, on occasion, any serotype can produce any of the syndromes. In general, more serious infections occur in infants, in adults over the age of 50, and in subjects with debilitating illnesses.

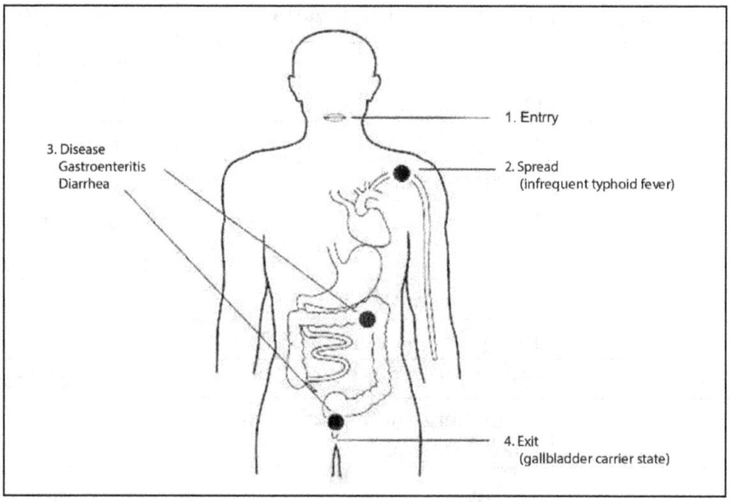

Figure: Pathogenesis of salmonellosis.

Most non-typhoidal salmonellae enter the body when contaminated food is ingested. Person-to-person spread of salmonellae also occurs. To be fully pathogenic, salmonellae must possess a variety of attributes called virulence factors. These include the ability to invade cells, a complete lipopolysaccharide coat, the ability to replicate intracellularly, and possibly the elaboration of toxin(s). After ingestion, the organisms colonize the ileum and colon, invade the intestinal epithelium, and proliferate within the epithelium and lymphoid follicles. The mechanism by which salmonellae invade the epithelium is partially understood and involves an initial binding to specific receptors on the epithelial cell surface followed by invasion. Invasion occurs by the organism inducing the enterocyte membrane to undergo "ruffling" and thereby to stimulate pinocytosis of the organisms. Invasion is dependent on rearrangement of the cell cytoskeleton and probably involves increases in cellular inositol phosphate and calcium. Attachment and invasion are under distinct genetic control and involve multiple genes in both chromosomes and plasmids.

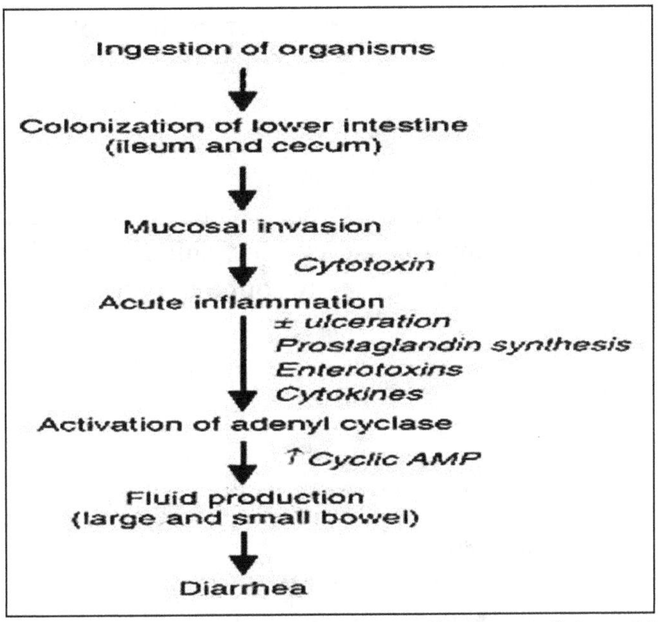

Figure: Scheme of the Pathogenesis of *Salmonella* enterocolitis and diarrhea.

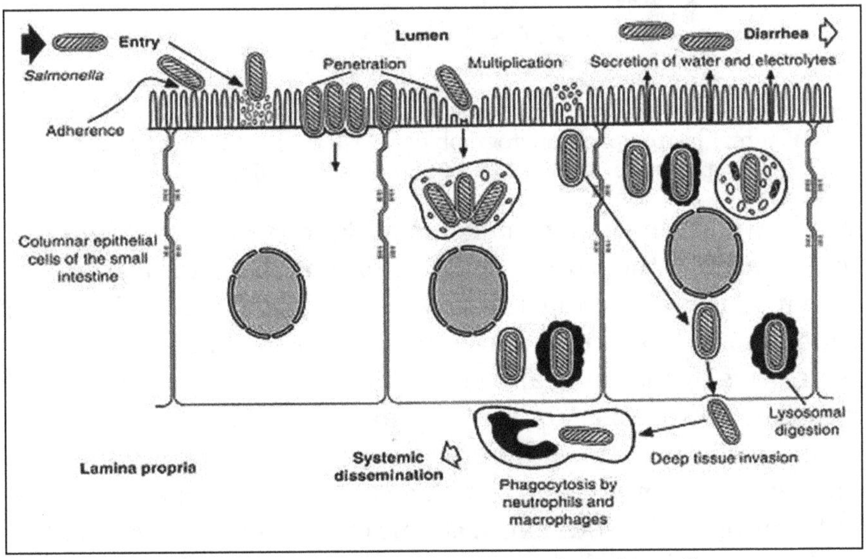

Figure: Invasion of intestinal mucosa by Salmonella.

After invading the epithelium, the organisms multiply intracellularly and then spread to mesenteric lymph nodes and throughout the body via the systemic circulation; they are taken up by the reticuloendothelial cells. The reticuloendothelial system confines and controls spread of the organism. However, depending on the serotype and the effectiveness of the host defenses against that serotype, some organisms may infect the liver, spleen, gallbladder, bones, meninges, and other organs. Fortunately, most serovars are killed promptly in extraintestinal sites, and the most common human Salmonella infection, gastroenteritis, remains confined to the intestine.

After invading the intestine, most salmonellae induce an acute inflammatory response, which can cause ulceration. They may elaborate cytotoxins that inhibit protein synthesis. Whether these cytotoxins contribute to the inflammatory response or to ulceration is not known. However, invasion of the mucosa causes the epithelial cells to synthesize and release various proinflammatory cytokines, including: IL-1, IL-6, IL-8, TNF-2, IFN-U, MCP-1, and GM-CSF. These evoke an acute inflammatory response and may also be responsible for damage to the intestine. Because of the intestinal inflammatory reaction, symptoms of inflammation such as fever, chills, abdominal pain, leukocytosis, and diarrhea are common. The stools may contain polymorphonuclear leukocytes, blood, and mucus.

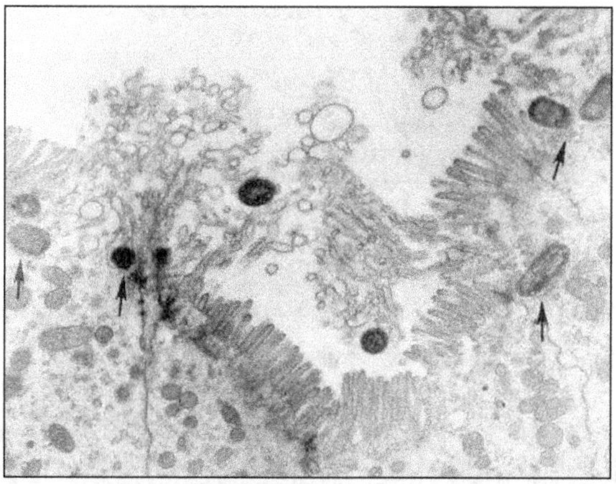

Figure: Electron photomicrograph demonstrating invasion of guinea pig ileal epithelial cells by Salmonella typhimurium.

Much is now known about the mechanisms of Salmonella gastroenteritis and diarrhea. Figure summarizes the pathogenesis of Salmonella enterocolitis and diarrhea. Only strains that penetrate the intestinal mucosa are associated with the appearance of an acute inflammatory reaction and diarrhea; the diarrhea is due to secretion of fluid and electrolytes by the small and large intestines. The mechanisms of secretion are unclear, but the secretion is not merely a manifestation of tissue destruction and ulceration. Salmonella penetrate the intestinal epithelial cells but, unlike Shigella and invasive E. coli, do not escape the phagosome. Thus, the extent of intercellular spread and ulceration of the epithelium is minimal. Salmonella escape from the basal side of epithelial cells into the lamina propria. Systemic spread of the organisms can occur, giving rise to enteric fever. Invasion of the intestinal mucosa is followed by activation of mucosal adenylate cyclase; the resultant increase in cyclic AMP induces secretion. The mechanism by which adenylate cyclase is stimulated is not understood; it may involve local production of prostaglandins or other components of the inflammatory reaction. In addition, Salmonella strains elaborate one or more enterotoxin-like substances which may stimulate intestinal secretion. However, the precise role of these toxins in the pathogenesis of Salmonella enterocolitis and diarrhea has not been established.

Host Defenses

Various host defenses are important in resisting intestinal colonization and invasion by Salmonella. Normal gastric acidity ($pH < 3.5$) is lethal to salmonellae. In healthy individuals, the number of ingested salmonellae is reduced in the stomach, so that fewer or no organisms enter the intestine. Normal small intestinal motility also protects the bowel by sweeping ingested salmonellae through quickly. The normal intestinal microflora protects against salmonellae, probably through anaerobes, which liberate short-chain fatty acids that are thought to be toxic to salmonellae. Alteration of the anaerobic intestinal flora by antibiotics renders the host more susceptible to salmonellosis. Secretory or mucosal antibodies also protect the intestine against salmonellae. Animal strains genetically resistant to intestinal invasion by salmonellae have been described. When these host defenses are absent or blunted, the host becomes more susceptible to salmonellosis; factors that render the host more susceptible to salmonellosis are listed in Table. For example, in AIDS, Salmonella infection is common, frequently persistent and bacteremic, and often resistant to even prolonged antibiotic treatment. Relapses are common. The role of host defenses in salmonellosis is extremely important, and much remains to be learned.

Table: Host Defenses against Salmonellae.

Host Defenses	Examples of Factors
Gastric Factors	Gastric acidity. Rate of gastric emptying.
Intestinal Factors	Intestinal motility. Normal intestinal flora. Mucus. Secretory antibodies. Genetic resistence to invasion.
Nonspecific and Other Possible Factors	Nutritional state. Lactoterrin. Gut reticuloendothelial cells Lysozyme.

Table: Factors Increasing Susceptibility to Salmonellosis.

Location or Factor	Specific Condition.
Stomach	Achlorhydria. Gastric Surgery.
Intestine	Antibiotic administration. Gastrountestinal surgery. Idiopathic inflammatory bowel desease.
Hemolytic	Especially sickle cell anemia and other hemoglobinopathies.
Impaired System Immunity	Carcinomatosis, luekemias, lymphomas. Diabetes mellitus, Immunosuppressive drugs, acquired immunodeficiency syndrome (AIDS), etc.

References

- Bacterial-sizes-shapes-arrangement: microbenotes.com, Retrieved 22, May 2020
- Bacterial-flagella-structure-importance-and-examples-of-flagellated-bacteria: microbeonline.com, Retrieved 13, July 2020
- Staphylococcus-aureus-overview-of-bacteriology-clinical-diseases-epidemiology-antibiotic-resistance, frontiers-in-i-staphylococcus-aureus-i: intechopen.com, Retrieved 12, Jan 2020

Chapter 3

Growth and Genetics of Bacteria

Bacterial genetics is the study of genetic materials of bacteria and their transfer during procreation. Through binary fission, bacteria divide into two daughter cells leading to bacterial reproduction. This chapter on bacterial genetics, bacterial growth and the various processes associated with the transfer of genetic material has been carefully written to provide an easy understanding of the topic.

Bacterial Genetics

Bacterial genetics is the study of the mechanisms of heritable information in bacteria, their chromosomes, plasmids, transposons and phages. Techniques that have enabled this discipline are culture in defined media, replica plating, mutagenesis, transformation, conjugation and transduction.

Beneficial mutations that develop in one bacterial cell can also be passed to related bacteria of different lineages through the process of horizontal transmission. There are three main forms of horizontal transmission used to spread genes between members of the same or different species: conjugation (bacteria-to-bacteria transfer), transduction (viral-mediated transfer), and transformation (free DNA transfer). These forms of genetic transfer can move plasmid , bacteriophage, or genomic DNA sequences. A plasmid is a small circle of DNA separate from the chromosome; a bacteriophage is a virus that reproduces in bacteria by injecting its DNA; the genome is the total DNA of the bacterial organism.

After transfer, the DNA molecules can exist in two forms, either as DNA molecules separate from the bacterial chromosome (an episome), or can become part of the bacterial chromosome. The study of basic mechanisms used by bacteria to exchange genes allowed scientists to develop many of the essential tools of modern molecular biology.

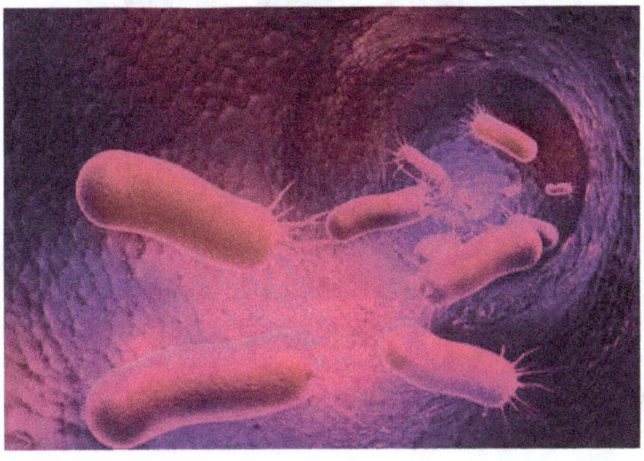

Conjugation

Bacterial conjugation refers to the transfer of DNA between bacterial cells that requires cell-to-cell contact. Joshua Lederberg and Edward Tatum first described conjugation in 1946 when they discovered the F factor (an episome) that can move between Escherichia coli cells. The F factor is one of the most well studied conjugative plasmids (plasmids are circular episomes) and is the most well studied conjugative system. There are many different conjugal plasmids carried by members of most bacterial species. Conjugal plasmids that carry antibiotic resistance genes are called R factors. The F factor and R factors usually exist as episomes and each carries functions that allow it to replicate its DNA and thus be inherited by the daughter cells after binary fission. However, conjugative plasmids also express transfer functions that allow the movement of DNA from a donor to a recipient cell; this is the process of conjugation.

The steps of bacterial conjugation are: mating pair formation, conjugal DNA synthesis, DNA transfer, and maturation. The main structure of the F factor that allows mating pair formation is the F pilus or sex pilus (a long thin fiber that extends from the bacterial cell surface). There are one to three pili expressed on an E. coli cell that carries the F factor, and one pilus will specifically interact with several molecules on the recipient cell surface (attachment). About twenty genes on the F factor are required to produce a functional pilus, but the structure is mainly made up of one protein , pilin. To bring the donor and recipient cell into close proximity, the F pilus retracts into the donor cell by removing pilin protein monomers from the base of the pilus to draw the bacterial cells together.

Once a stable mating pair is formed, a specialized form of DNA replication starts. Conjugal DNA synthesis produces a single-stranded copy of the F factor DNA (as opposed to a double-stranded DNA that is formed by normal replication). This DNA strand is transferred into the recipient cell. Once in the recipient cell, the single-stranded copy of the F plasmid DNA is copied to make a double-stranded DNA molecule, which then forms a mature circular plasmid. At the end of conjugation the mating pair is broken and both the donor and the recipient cells carry an identical episomal copy of the F factor. All of the approximately one hundred genes carried on the F factor can now be expressed by the recipient cell and will be inherited by its offspring.

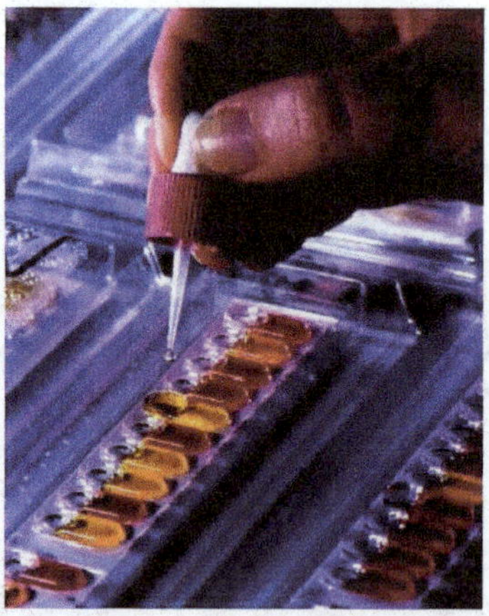

A laboratory technician performing an Analytical Profile Index (API) test on bacteria.

In addition to transferring itself, the F factor can also transfer chromosomal genes between a donor and recipient cell. The F factor can be found inserted (integrated) into the bacterial chromosome at many locations in a small fraction of bacterial cells. An integrated F factor is replicated along with the rest of the chromosome and inherited by offspring along with the rest of the chromosome. When a mating pair is formed between the donor cell carrying an integrated F factor and a recipient cell, DNA transfer occurs as it does for the episomal F factor, but now the chromosomal sequences adjacent to the integrated F factor are transferred into the recipient. Since these DNA sequences encode bacterial genes, they can recombine with the same genes in the recipient. If the donor gene has minor changes in DNA sequence from the recipient gene, the different sequence can be incorporated into the recipient gene and inherited by the recipient cell's offspring. Donor cells that have an integrated copy of the F factor are called Hfr strains (High frequency of recombination).

Transduction

The second way that DNA is transferred between bacterial cells is through a phage particle in the process of transduction. Joshua Lederberg and Norton Zinder first discovered transduction in 1956. When phage inject their DNA into a recipient cell, a process occurs that produces new bacteriophage particles and kills the host cell (lytic growth). Some phage do not always kill the host cell (temperate phage), but instead can be inherited by daughter host cells. Therefore acquisition of a so-called temperate "prophage" by a recipient cell is a form of transduction. Many phage also have the ability to transfer chromosomal or plasmid genes between bacterial cells. During generalized transduction any gene can be transferred from a donor cell to a recipient cell. Generalized transducing phage are produced when a phage packages bacterial genes into its capsid (protein envelope) instead of its own DNA. When a phage particle carrying bacterial chromosomal genes attaches to a recipient cell, the DNA is injected into the cytoplasm where it can recombine with a homologous DNA sequences.

Some bacteriophage can pick up a subset of chromosomal genes and transfer them to other bacteria. This process is called specialized transduction since only a limited set of chromosomal genes can be transferred between bacterial cells.

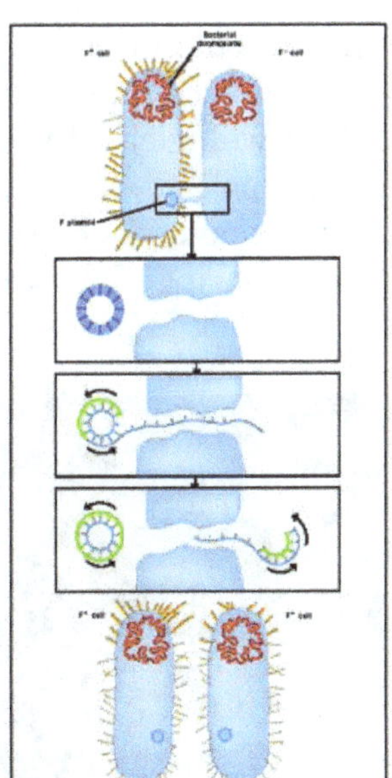

Bacterial conjugation. The bacterium on the left passes a copy of the F plasmid to the bacterium on the right, converting it from an F - cell to an F + cell.

The third main way that bacteria exchange DNA is called DNA transformation. Some bacteria have evolved systems that transport free DNA from the outside of the bacterial cell into the cytoplasm. These bacterial are called "naturally competent" for DNA transformation. Natural DNA transformation of *Streptococcus pneumonaiae* provided the first proof that DNA encoded the genetic material in experiments by Oswald Avery and colleagues. Some other naturally competent bacteria include *Bacillus subtilis, Haemophilus influenzae,* and *Neisseria gonorrhoeae.* Other bacterial species such as *E. coli* are not naturally competent for DNA transformation. Scientists have devised many ways to physically or chemically force noncompetant bacteria to take up DNA. These methods of artificial DNA transformation form the basis of plasmid cloning in molecular biology.

Most naturally competent bacteria regulate transformation competence so that they only take up DNA into their cells when there is a high density of cells in the environment. The ability to sense how many other cells are in an area is called quorum sensing. Bacteria that are naturally competent for DNA transformation express ten to twenty proteins that form a structure that spans the bacterial cell envelope. In some bacteria this structure also is required to form a particular type of pilus different than the F factor pilus. Other bacteria express similar structures that are involved in secreting proteins into the exterior medium (Type II secretion). Therefore, it appears that DNA transformation and protein secretion have evolved together.

During natural DNA transformation, doubled-stranded DNA is bound to the recipient cell surface by a protein receptor. One strand of the DNA is transported through the cell envelope, where it can recombine with similar sequences present in the recipient cell. If the DNA taken up is not homologous to genes already present in the cell, the DNA is usually broken down and the nucleotides released are used to synthesize new DNA during normal replication. This observation has led to the speculation that DNA transformation competence may have originally evolved to allow the acquisition of nucleic acids for food.

The source of DNA for transformation is thought to be DNA released from other cells in the same population. Most naturally competent bacteria spontaneously break apart by expressing enzymes that break the cell wall. Autolysis will release the genomic DNA into the environment where it will be available for DNA transformation. Of course, this results in the death of some cells in the population, but usually not large numbers of cells. It appears that losing a few cells from the population is counterbalanced by having the possibility of gaining new traits by DNA transformation.

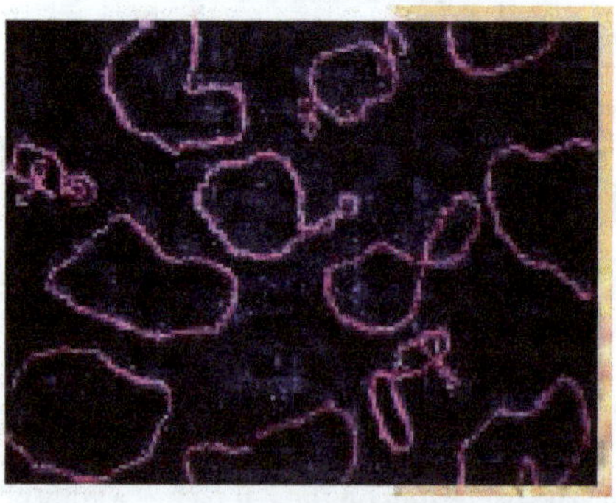

A scanning electron micrograph of bacterial DNA plasmids.

Bacterial Growth

Bacterial growth is finely tuned to cellular state and environmental input. Growth not only determines the fate of a single cell, but also of the community, and both depends on and is regulated by other cells, including kin cells, other microorganisms and host cells. What governs bacterial growth is a classic question of microbiology with a rich history of research, and modern tools and approaches have invigorated the field.

Bacteria replicate by binary fission, a process by which one bacterium splits into two. Therefore, bacteria increase their numbers by geometric progression whereby their population doubles every generation time. Generation time is the time it takes for a population of bacteria to double in number. For many common bacteria, the generation time is quite short, 20-60 minutes under optimum conditions. For most common pathogens in the body, the generation time is probably closer to 5-10 hours. Because bacteria grow by geometric progression and most have a short generation time, they can astronomically increase their number in a short period of time.

The relationship between the number of bacteria in a population at a given time (Nt), the original number of bacterial cells in the population (No), and the number of divisions those bacteria have undergone during that time (n) can be expressed by the following equation:

$$N_t = N_o \times 2^n$$

For example, Escherichia coli, under optimum conditions, has a generation time of 20 minutes. If one started with only 10 E. coli (No = 10) and allowed them to grow for 12 hours (n = 36; with a generation time of 20 minutes they would divide 3 times in one hour and 36 times in 12 hours), then plugging the numbers in the formula, the number of bacteria after 12 hours (Nt) would be,

$$10 \times 2^{36} = N_t = 687{,}194{,}767{,}360 \text{E.coli}$$

In general it is thought that during DNA replication, each strand of the replicating bacterial DNA attaches to proteins at what will become the cell division plane. For example, Par proteins function to separate bacterial chromosomes to opposite poles of the cell during cell division. They bind to the origin of replication of the DNA and physically pull or push the chromosomes apart, similar to the mitotic apparatus of eukaryotic cells.

Physical Conditions Required for Growth

The physical environment in which the organisms will grow best. They exhibit diverse responsive to physical conditions such as Temperature, Gaseous conditions, and pH.

Temperature

The best temperature for bacterial growth is different. If some bacteria grows in artic oceans of hot springs, it is not surprising matter. The optimum temperature of bacteria is not constant. In human GUT some bacteria grows very well at human body temperature 37°C. But plant bacteria are not able to survive at those temperatures.

All processes of growth are dependent on chemical reactions and since the rates of those reactions are influenced by temperature. The temperature that allows for most rapid growth during a short period of time (12 to 24 hours) is known as the “Optimum temperature”.

On the basis of their temperature relationships, Bacteria are divided into three main groups:

Psychrophiles

- These are able to grow at 0°C (or) lower, though they grow best at a higher temperature.
- Many microbiologists restrict the term Psychrophile to organisms that can grow at 0°C but have an optimum temperature of 15°C (or) lower and maximum temperature of about 20°C.
- Facultative psychrophile organisms are able to grow at 0°C; and temperature in the range of about 20 to 30°C.

Mesophiles

- These grow best within a temperature range of approximately 25 to 40°C.E.g.: Pathogenic bacteria for humans and Warm-blooded animals.

Thermophiles

- These grow best at temperatures above 45°C.
- The growth range of many thermopiles extends into the mesophilic region; these species are designated “Facultative thermophiles”.

Oxygen Toxicity

- Oxygen is both beneficial and poisonous to living organisms.
- It is beneficial its strong oxidizing ability makes it an excellent terminal electron acceptor for the energy-yielding process known as “Respiration”.
- Oxygen is also a toxic substance.

The following factors are among those that have been implicated in oxygen toxicity.

Oxygen Inactivation of Enzymes

Molecular oxygen can directly oxidize certain essential reduced groups such as thiol (-SH) groups (or) enzymes, results in enzyme inactivation.

Eg: Nitrogenase (nitrogen fixation enzymes) destroyed by even small amounts of oxygen.

Damage Due to Toxic Derivatives of Oxygen

Various cellular enzymes catalyze chemical reactions involving molecular oxygen; these reactions can result in the addition of a single electron to an O_2 molecule, by the result “Superoxide radical”(O_2^-).

$O_2 + e^- \rightarrow O_{2-}$

These superoxide radicals can inactivated vital cell components. Recent studies suggest the production of even more toxic substances by detrimental action of superoxide radical & protons form "Hydrogen peroxide (H_2O_2)" and "Hydroxyl radicals(OH.)".

$2\,O_2{-} + 2H^+ \rightarrow O_2 + H_2O_2$

$O_2{-} + H_2O_2 \rightarrow O_2 + OH^- + OH.$

This superoxide radical will eliminate by greatly increasing the enzyme "Superoxide dismutase". These hydrogen peroxide produced by this reaction can, in turn, be dissipated by "Catalase" and "Peroxidase" enzymes.

$2\,H_2O_2 \rightarrow 2\,H_2O + O_2$

H_2O_2 + Reduced Substrate -> 2 H_2O + Oxidized Substrate

Elimination of either super-oxide radicals (or) hydrogen peroxide can prevent the formation of the highly dangerous hydroxyl radicals.

Acidity (or) Alkalinity (pH)

For most Bacteria, the optimum pH for growth lies between 6.5 and 7.5. Few Bacteria prefer more extreme pH values for growth. For example, "Thiobacillus thiooxidants" has optimum pH of 2.0 to 3.5 and can grow in a range between pH 0.5 and 6.0.

When Bacteria are cultivated in a medium originally adjusted to a given pH, for example, 7.0, it is very likely that this pH will change as a result of the chemical activities of the organisms.

Incorporating a buffer into the medium can prevent a radical shift in pH. A buffer is a mixture of a weak acid and its conjugate base (e.g.: Acetic acid (CH_3COOH) and acetate (CH_3COO^-). Phosphate buffer (combination of $H_2PO_4^-$ and H $PO4^{-2}$) having a pKa of 6.8, is widely used in bacteriological media.

Some large fermentation apparatus are equipped with automatic controls that maintain a constant pH.

Types of Bacterial Growth

Bacterial growth refers to the increase in cell number or cell population of bacteria. It is brought about by cell multiplication.

Methods of Bacterial Growth

Bacteria grow by the following methods of reproduction:

1. Binary fission
2. Budding
3. Filamentation
4. Sporulation

Binary Fission

Binary fission refers to the division of the parent cell into two daughter cells. During binary fission, both cytoplasm and nuclei divide.

E.g.: Bacillus, streptococcus, etc.

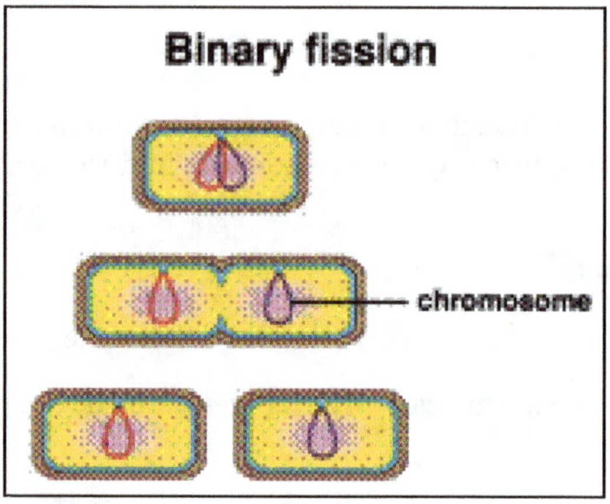

Binary fission in a). Bacillus b). Prosthecobacter.

Budding

In budding, the bacterial cell produces a small projection on its surface. It is called bud. It increases in size and separates into a daughter cell.

E.g.: Rhodopseudomonas

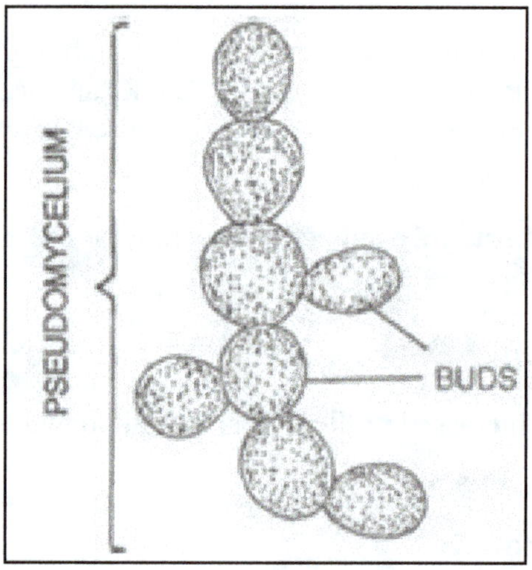

Filamentation

Filamentous Bacteria, the filaments break into fragments and each fragment grows into a daughter filament. This process of growth is called filamentation or fragmentation. E.g.: Nacardia.

Sporulation

Certain Bacteria grow by producing spores. The hypha at tip produces many spores. The spores separate and develop into new colonies.

E.g.: Streptomyces

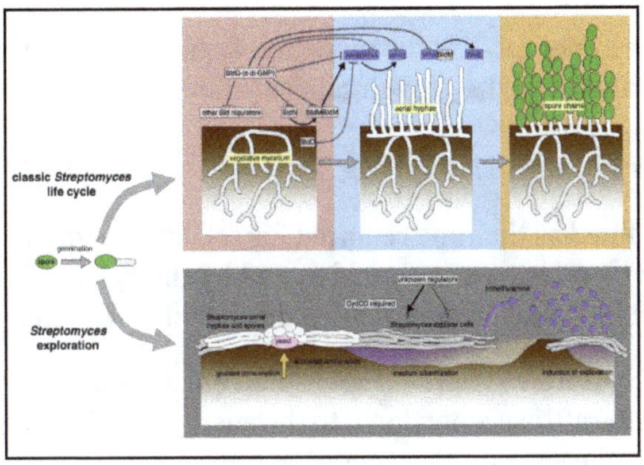

Bacterial Growth Rate

During growth, a cell divides into two and the two then divide into four unto 8 soon. In this way, the cells grow by geometrical progression.

1 -> 2 -> 4 -> 8 -> 16 -> 32 -> 64 ->

The time required by a cell to divide or the time required for the population to double is called generation time. The generation time varies from species to species and nutrients, temperature, etc affect it. The generation time for E.coli in milk at 37°C is 12 minutes.

Number of cells (N) derived from a single cell after 'n' generations will be,

$N=2^n$

Similarly, after 'n' generations, the number of cells derived from an initial cell population of NO will be,

$N=2^n NO$

The rate of growth of a culture,

Different parameters are used to express the rate of growth in bacterial culture. They are:

Growth rate constant denoted as 'k',

Mean doubling time or generation time denoted as 'g'.

The growth rate of a cell refers to how rapidly a cell increases in mass. The rate of increase in bacteria at any particular time is proportional to the number or mass of Bacteria present at that time. The bacterial growth equation is,

The Rate Increase of Cells = k (No. of Cells or Mass of Cells)

Here 'k' is the proportionality constant and is an index of the rate of growth; 'k' also to the amount of that cellular component. In mathematical terms,

$$\frac{dN}{dt} = kN; \quad \frac{dX}{dt} = kX; \; and \; \frac{dZ}{dt} = kZ$$

Here N is the number of cells per ml, X is the mass of cells/ml, Z is the amount of any cellular component/ml, 't' is the time and 'K' is the growth rate constant.

Generation Time (or) Mean Doubling Time

The generation time is defined as the time required for cell components of the culture or the number (or) mass to increase by a factor of 2. It is denoted by the symbol 'g'. Generation time is also referred to the time that a bacterial cell takes to go through its life cycle.

Generation time is usually shorter for prokaryotes than eukaryotes, a shorter for smaller than for larger cells since the growth rate is proportional to the energy metabolism of the cell. The faster the cell metabolizes nutrients, the shorter it's generation time.

Growth Curve

The increase in the cell size and cell mass during the development of an organism is termed as growth. It is the unique characteristics of all organisms. The organism must require certain basic parameters for their energy generation and cellular biosynthesis. The growth of the organism is affected by both physical and Nutritional factors. The physical factors include the pH, temperature, Osmotic pressure, Hydrostatic pressure, and Moisture content of the medium in which the organism is growing. The nutritional factors include the amount of Carbon, nitrogen, Sulphur, phosphorous, and other trace elements provided in the growth medium. Bacteria are unicellular (single cell) organisms. When the bacteria reach a certain size, they divide by binary fission, in which the one cell divides into two, two into four and continue the process in a geometric fashion. The bacterium is then known to be in an actively growing phase. To study the bacterial growth population, the viable cells of the bacterium should be inoculated on to the sterile broth and incubated under optimal growth conditions. The bacterium starts utilising the components of the media and it will increase in its size and cellular mass. The dynamics of the bacterial growth can be studied by plotting the cell growth (absorbance) versus the incubation time or log of cell number versus time. The curve thus obtained is a sigmoid curve and is known as a standard growth curve. The increase in the cell mass of the organism is measured by using the Spectrophotometer. The Spectrophotometer measures the turbidity or Optical density which is the measure of the amount of light absorbed by a bacterial suspension. The degree of turbidity in the broth culture is directly related to the number of microorganism present, either viable or dead cells, and is a convenient and rapid method of measuring cell growth rate of an organism. Thus the increasing the turbidity of the broth medium indicates increase of the microbial cell mass. The amount of transmitted light through turbid broth decreases with subsequent increase in the absorbance value.

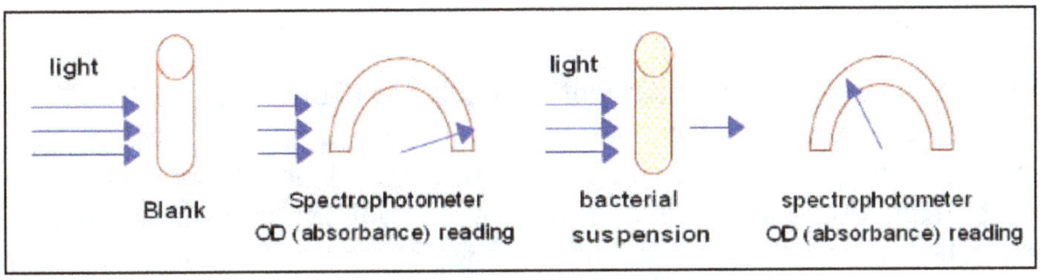

Figure: Absorbance reading of bacterial suspension.

The growth curve has four distinct phases.

1. Lag phase:

When a microorganism is introduced into the fresh medium, it takes some time to adjust with the new environment. This phase is termed as Lag phase, in which cellular metabolism is accelerated, cells are increasing in size, but the bacteria are not able to replicate and therefore no increase in cell mass. The length of the lag phase depends directly on the previous growth condition of the organism. When the microorganism growing in a rich medium is inoculated into nutritionally poor medium, the organism will take more time to adapt with the new environment. The organism will start synthesising the necessary proteins, co-enzymes and vitamins needed for their growth and hence there will be a subsequent increase in the lag phase. Similarly when an organism from a nutritionally poor medium is added to a nutritionally rich medium, the organism can easily adapt to the environment, it can start the cell division without any delay, and therefore will have less lag phase it may be absent.

2. Exponential or Logarithmic (log) phase:

During this phase, the microorganisms are in a rapidly growing and dividing state. Their metabolic activity increases and the organism begin the DNA replication by binary fission at a constant rate. The growth medium is exploited at the maximal rate, the culture reaches the maximum growth rate and the number of bacteria increases logarithmically (exponentially) and finally the single cell divide into two, which replicate into four, eight, sixteen, thirty two and so on (That is 20, 21, 22, 23.........2n, n is the number of generations) This will result in a balanced growth. The time taken by the bacteria to double in number during a specified time period is known as the generation time. The generation time tends to vary with different organisms. E.coli divides in every 20 minutes, hence its generation time is 20 minutes, and for Staphylococcus aureus it is 30 minutes.

3. Stationary phase:

As the bacterial population continues to grow, all the nutrients in the growth medium are used up by the microorganism for their rapid multiplication. This result in the accumulation of waste materials, toxic metabolites and inhibitory compounds such as antibiotics in the medium. This shifts the conditions of the medium such as pH and temperature, thereby creating an unfavourable environment for the bacterial growth. The reproduction rate will slow down, the cells undergoing division is equal to the number of cell death, and finally bacterium stops its division completely. The cell number is not increased and thus the growth rate is stabilised. If a cell taken from the stationary phase is introduced into a fresh medium, the cell can easily move on the exponential phase and is able to perform its metabolic activities as usual.

4. Decline or Death phase:

The depletion of nutrients and the subsequent accumulation of metabolic waste products and other toxic materials in the media will facilitates the bacterium to move on to the Death phase. During this, the bacterium completely loses its ability to reproduce. Individual bacteria begin to die due to the unfavourable conditions and the death is rapid and at uniform rate. The number of dead cells exceeds the number of live cells. Some organisms which can resist this condition can survive in the environment by producing endospores.

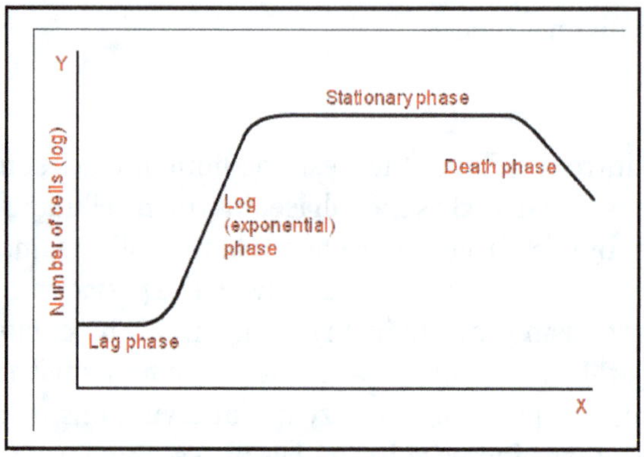

Figure: Different phases of growth of a bacteria.

Calculation

The generation time can be calculated from the growth curve.

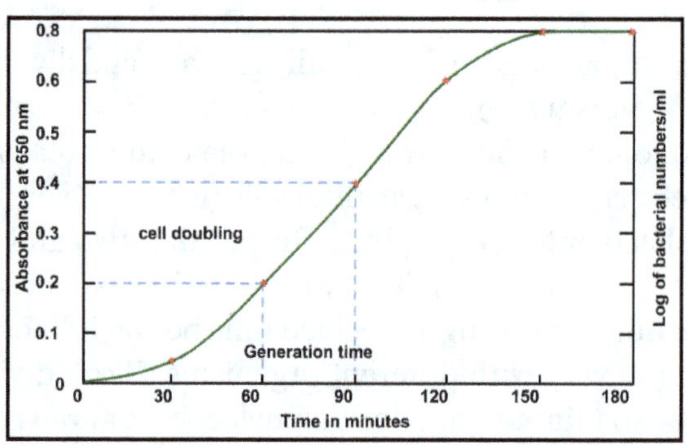

Figure: Calculation of generation time.

The exactly doubled points from the absorbance readings were taken and, the points were extrapolated to meet the respective time axis.

Generation Time = (Time in minutes to obtain the absorbance 0.4) – (Time in minutes to obtain the absorbance 0.2)

= 90-60

= 30 minutes

Let No = the initial population number.

Nt = population at time t

N = the number of generations in time t

Therefore,

$$Nt = No \times 2^n$$

$$\log Nt = \log No + n \log 2$$

Therefore,

$$n = (\log Nt - \log No) / \log 2$$

$$n = (\log Nt - \log No) / 0.301$$

The growth rate can be expressed in terms of mean growth rate constant (k), the number of generations per unit time.

$$k = n / t$$

$$k = (\log Nt - \log No) / (0.301 \times t)$$

Mean generation time or mean doubling time (g), is the time taken to double its size.

Therefore,

$$Nt = 2No$$

Substituting equation $Nt = 2No$ in equation $k = (\log Nt - \log No) / (0.301 \times t)$

$$k = (\log Nt - \log No) / (0.301 \times t)$$

$$= (\log 2No - \log No) / (0.301 \times t)$$

$= \log 2 + (\log No - \log No) / 0.301\ g$ (Since the population doubles t= g)

Therefore,

$$k = 1 / g$$

Mean growth rate constant, $k = 1 / g$

Mean generation time, $g = 1 / k$

Measuring Bacterial Growth

Estimating the number of bacterial cells in a sample, known as a bacterial count, is a common task performed by microbiologists. The number of bacteria in a clinical sample serves as an indication of the extent of an infection. Quality control of drinking water, food, medication, and even cosmetics relies on estimates of bacterial counts to detect contamination and prevent the spread of disease. Two major approaches are used to measure cell number. The direct methods involve counting cells, whereas the indirect methods depend on the measurement of cell presence or activity without actually counting individual cells. Both direct and indirect methods have advantages and disadvantages for specific applications.

Direct Cell Count

Direct cell count refers to counting the cells in a liquid culture or colonies on a plate. It is a direct way of estimating how many organisms are present in a sample. Let's look first at a simple and fast method that requires only a specialized slide and a compound microscope.

The simplest way to count bacteria is called the direct microscopic cell count, which involves transferring a known volume of a culture to a calibrated slide and counting the cells under a light microscope. The calibrated slide is called a Petroff-Hausser chamber and is similar to a hemocytometer used to count red blood cells. The central area of the counting chamber is etched into squares of various sizes. A sample of the culture suspension is added to the chamber under a coverslip that is placed at a specific height from the surface of the grid. It is possible to estimate the concentration of cells in the original sample by counting individual cells in a number of squares and determining the volume of the sample observed. The area of the squares and the height at which the coverslip is positioned are specified for the chamber. The concentration must be corrected for dilution if the sample was diluted before enumeration.

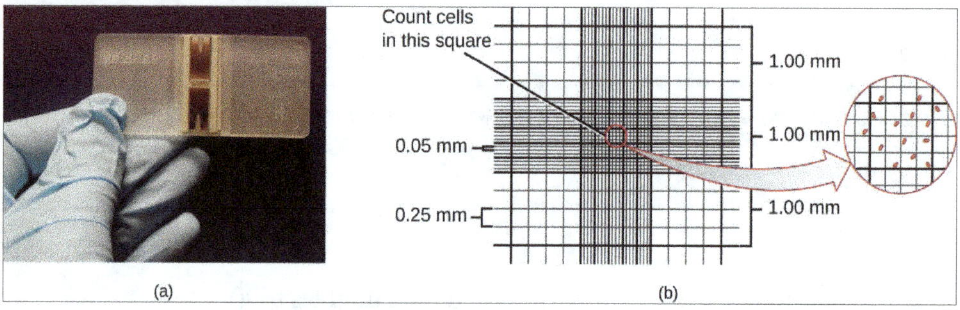

Figure: (a) A Petroff-Hausser chamber is a special slide designed for counting the bacterial cells in a measured volume of a sample. A grid is etched on the slide to facilitate precision in counting. (b) This diagram illustrates the grid of a Petroff-Hausser chamber, which is made up of squares of known areas. The enlarged view shows the square within which bacteria (red cells) are counted. If the coverslip is 0.2 mm above the grid and the square has an area of 0.04 mm2, then the volume is 0.008 mm3, or 0.000008 mL. Since there are 10 cells inside the square, the density of bacteria is 10 cells/0.000008 mL, which equates to 1,250,000 cells/mL.

Cells in several small squares must be counted and the average taken to obtain a reliable measurement. The advantages of the chamber are that the method is easy to use, relatively fast, and inexpensive. On the downside, the counting chamber does not work well with dilute cultures because there may not be enough cells to count.

Using a counting chamber does not necessarily yield an accurate count of the number of live cells because it is not always possible to distinguish between live cells, dead cells, and debris of the same size under the microscope. However, newly developed fluorescence staining techniques make it possible to distinguish viable and dead bacteria. These viability stains (or live stains) bind to nucleic acids, but the primary and secondary stains differ in their ability to cross the cytoplasmic membrane. The primary stain, which fluoresces green, can penetrate intact cytoplasmic membranes, staining both live and dead cells. The secondary stain, which fluoresces red, can stain a cell only if the cytoplasmic membrane is considerably damaged. Thus, live cells fluoresce green because they only absorb the green stain, whereas dead cells appear red because the red stain displaces the green stain on their nucleic acids.

Another technique uses an electronic cell counting device (Coulter counter) to detect and count the changes in electrical resistance in a saline solution. A glass tube with a small opening is immersed in an electrolyte solution. A first electrode is suspended in the glass tube. A second electrode is located outside of the tube. As cells are drawn through the small aperture in the glass tube, they briefly change the resistance measured between the two electrodes and the change is recorded by an electronic sensor; each resistance change represents a cell. The method is rapid and accurate within a range of concentrations; however, if the culture is too concentrated, more than one cell may pass through the aperture at any given time and skew the results. This method also does not differentiate between live and dead cells.

Direct counts provide an estimate of the total number of cells in a sample. However, in many situations, it is important to know the number of live, or viable, cells. Counts of live cells are needed when assessing the extent of an infection, the effectiveness of antimicrobial compounds and medication, or contamination of food and water.

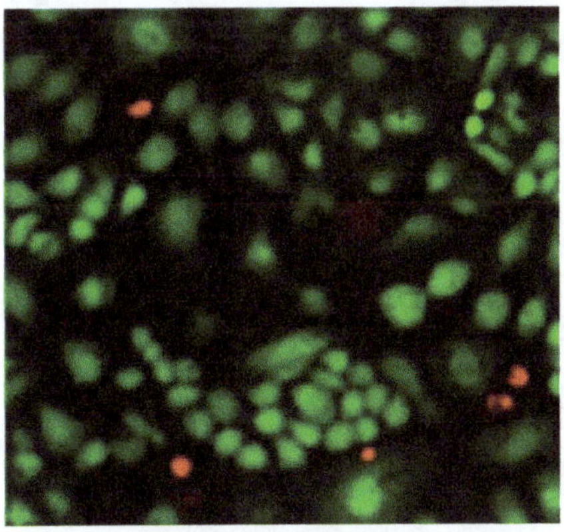

Figure: Fluorescence staining can be used to differentiate between viable and dead bacterial cells in a sample for purposes of counting. Viable cells are stained green, whereas dead cells are stained red.

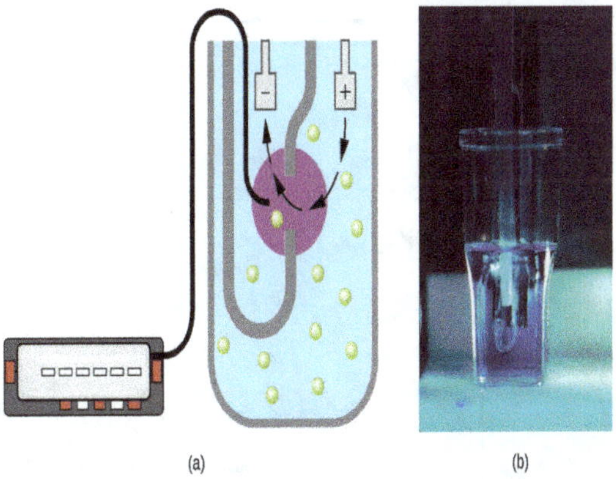

Figure: A Coulter counter is an electronic device that counts cells. It measures the change in resistance in an electrolyte solution that takes place when a cell passes through a small opening in the inside container wall. A detector automatically counts the number of cells passing through the opening.

Plate Count

The viable plate count, or simply plate count, is a count of viable or live cells. It is based on the principle that viable cells replicate and give rise to visible colonies when incubated under suitable conditions for the specimen. The results are usually expressed as colony-forming units per milliliter (CFU/mL) rather than cells per milliliter because more than one cell may have landed on the same spot to give rise to a single colony. Furthermore, samples of bacteria that grow in clusters or chains are difficult to disperse and a single colony may represent several cells. Some cells are described as viable but nonculturable and will not form colonies on solid media. For all these reasons, the viable plate count is considered a low estimate of the actual number of live cells. These limitations do not detract from the usefulness of the method, which provides estimates of live bacterial numbers.

Microbiologists typically count plates with 30–300 colonies. Samples with too few colonies (<30) do not give statistically reliable numbers, and overcrowded plates (>300 colonies) make it difficult to accurately count individual colonies. Also, counts in this range minimize occurrences of more than one bacterial cell forming a single colony. Thus, the calculated CFU is closer to the true number of live bacteria in the population.

There are two common approaches to inoculating plates for viable counts: the pour plate and the spread plate methods. Although the final inoculation procedure differs between these two methods, they both start with a serial dilution of the culture.

Serial Dilution

The serial dilution of a culture is an important first step before proceeding to either the pour plate or spread plate method. The goal of the serial dilution process is to obtain plates with CFUs in the range of 30–300, and the process usually involves several dilutions in multiples of 10 to simplify calculation. The number of serial dilutions is chosen according to a preliminary estimate of the culture density. Figure 10 illustrates the serial dilution method.

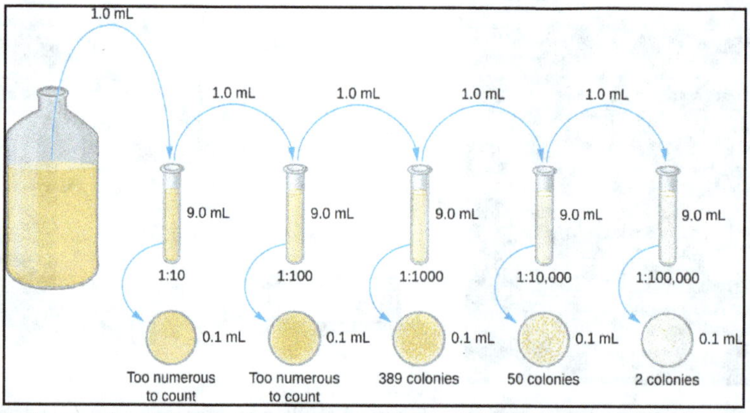

Figure: Serial dilution involves diluting a fixed volume of cells mixed with dilution solution using the previous dilution as an inoculum. The result is dilution of the original culture by an exponentially growing factor.

A fixed volume of the original culture, 1.0 mL, is added to and thoroughly mixed with the first dilution tube solution, which contains 9.0 mL of sterile broth. This step represents a dilution factor of 10, or 1:10, compared with the original culture. From this first dilution, the same volume, 1.0 mL, is withdrawn and mixed with a fresh tube of 9.0 mL of dilution solution. The dilution factor is now 1:100 compared with the original culture. This process continues until a series of dilutions is produced that will bracket the desired cell concentration for accurate counting. From each tube, a sample is plated on solid medium using either the pour plate method or the spread plate method. The plates are incubated until colonies appear. Two to three plates are usually prepared from each dilution and the numbers of colonies counted on each plate are averaged. In all cases, thorough mixing of samples with the dilution medium (to ensure the cell distribution in the tube is random) is paramount to obtaining reliable results.

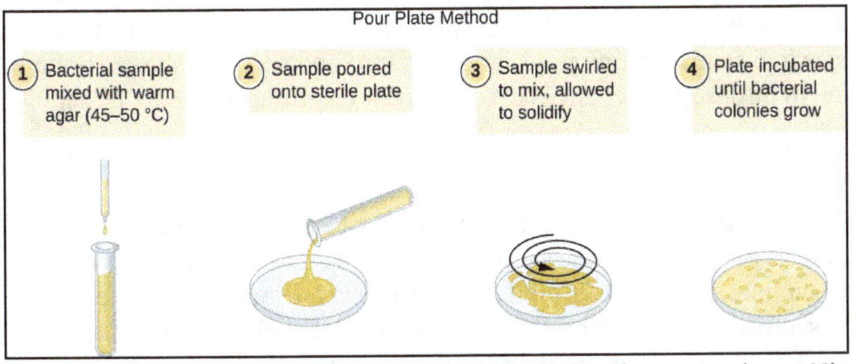

Figure: In the pour plate method of cell counting, the sample is mixed in liquid warm agar (45–50°C) poured into a sterile Petri dish and further mixed by swirling. This process is repeated for each serial dilution prepared. The resulting colonies are counted and provide an estimate of the number of cells in the original volume sampled.

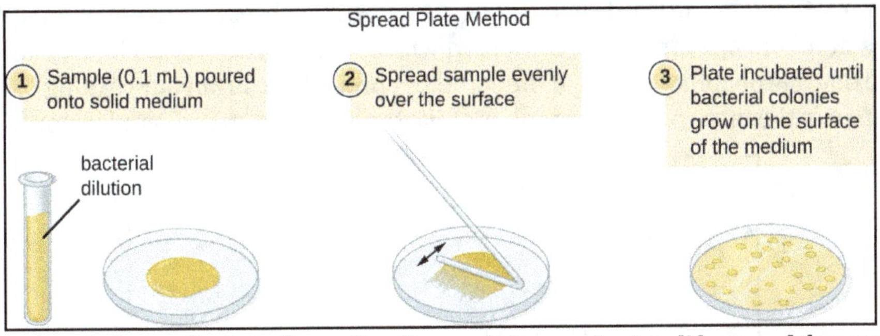

Figure: In the spread plate method of cell counting, the sample is poured onto solid agar and then spread using a sterile spreader. This process is repeated for each serial dilution prepared. The resulting colonies are counted and provide an estimate of the number of cells in the original volume samples.

The dilution factor is used to calculate the number of cells in the original cell culture. In our example, an average of 50 colonies was counted on the plates obtained from the 1:10,000 dilution. Because only 0.1 mL of suspension was pipetted on the plate, the multiplier required to reconstitute the original concentration is 10 × 10,000. The number of CFU per mL is equal to 50 × 100 × 10,000 = 5,000,000. The number of bacteria in the culture is estimated as 5 million cells/mL. The colony count obtained from the 1:1000 dilution was 389, well below the expected 500 for a 10-fold difference in dilutions. This highlights the issue of inaccuracy when colony counts are greater than 300 and more than one bacterial cell grows into a single colony.

A very dilute sample—drinking water, for example—may not contain enough organisms to use either of the plate count methods described. In such cases, the original sample must be concentrated rather than diluted before plating. This can be accomplished using a modification of the plate count technique called the membrane filtration technique. Known volumes are vacuum-filtered aseptically through a membrane with a pore size small enough to trap microorganisms. The membrane is transferred to a Petri plate containing an appropriate growth medium. Colonies are counted after incubation. Calculation of the cell density is made by dividing the cell count by the volume of filtered liquid.

Most Probable Number

The number of microorganisms in dilute samples is usually too low to be detected by the plate count methods described thus far. For these specimens, microbiologists routinely use the most probable number (MPN) method, a statistical procedure for estimating of the number of viable microorganisms in a sample. Often used for water and food samples, the MPN method evaluates detectable growth by observing changes in turbidity or color due to metabolic activity.

A typical application of MPN method is the estimation of the number of coliforms in a sample of pond water. Coliforms are gram-negative rod bacteria that ferment lactose. The presence of coliforms in water is considered a sign of contamination by fecal matter. For the method illustrated in figure, a series of three dilutions of the water sample is tested by inoculating five lactose broth tubes with 10 mL of sample, five lactose broth tubes with 1 mL of sample, and five lactose broth tubes with 0.1 mL of sample. The lactose broth tubes contain a pH indicator that changes color from red to yellow when the lactose is fermented. After inoculation and incubation, the tubes are examined for an indication of coliform growth by a color change in media from red to yellow. The first set of tubes (10-mL sample) showed growth in all the tubes; the second set of tubes (1 mL) showed growth in two tubes out of five; in the third set of tubes, no growth is observed in any of the tubes (0.1-mL dilution). The numbers 5, 2, and 0 are compared with Mathematical Basics figure, which has been constructed using a probability model of the sampling procedure. From our reading of the table, we conclude that 49 is the most probable number of bacteria per 100 mL of pond water.

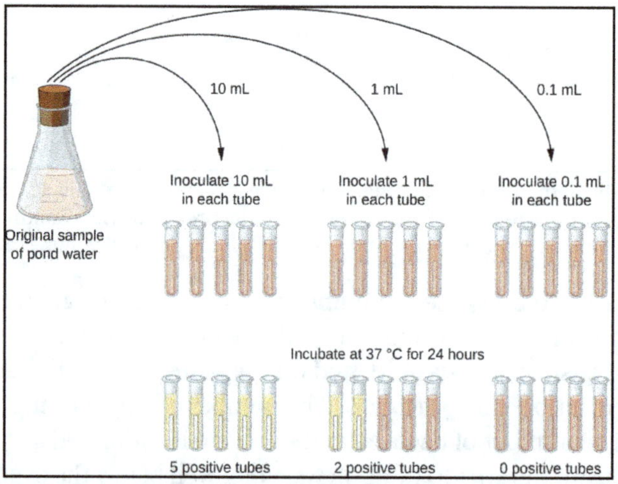

Figure: In the most probable number method, sets of five lactose broth tubes are inoculated with three different volumes of pond water: 10 mL, 1 mL, and 0.1mL. Bacterial growth is assessed through a change in the color of the broth from red to yellow as lactose is fermented.

Indirect Cell Counts

Besides direct methods of counting cells, other methods, based on an indirect detection of cell density, are commonly used to estimate and compare cell densities in a culture. The foremost approach is to measure the turbidity (cloudiness) of a sample of bacteria in a liquid suspension. The laboratory instrument used to measure turbidity is called a spectrophotometer. In a spectrophotometer, a light beam is transmitted through a bacterial suspension, the light passing through the suspension is measured by a detector, and the amount of light passing through the sample and reaching the detector is converted to either percent transmission or a logarithmic value called absorbance (optical density). As the numbers of bacteria in a suspension increase, the turbidity also increases and causes less light to reach the detector. The decrease in light passing through the sample and reaching the detector is associated with a decrease in percent transmission and increase in absorbance measured by the spectrophotometer.

Measuring turbidity is a fast method to estimate cell density as long as there are enough cells in a sample to produce turbidity. It is possible to correlate turbidity readings to the actual number of cells by performing a viable plate count of samples taken from cultures having a range of absorbance values. Using these values, a calibration curve is generated by plotting turbidity as a function of cell density. Once the calibration curve has been produced, it can be used to estimate cell counts for all samples obtained or cultured under similar conditions and with densities within the range of values used to construct the curve.

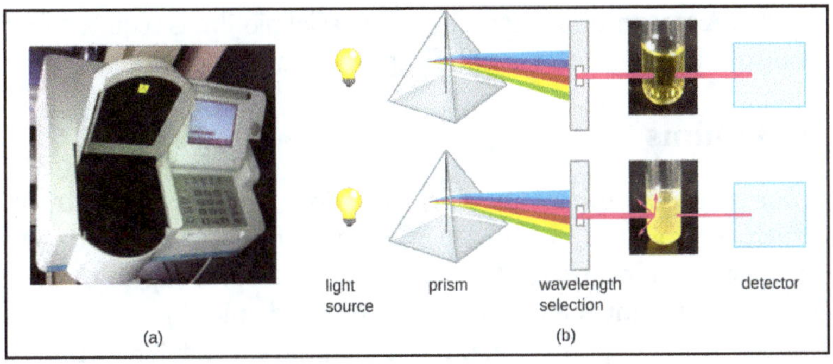

Figure: (a) A spectrophotometer is commonly used to measure the turbidity of a bacterial cell suspension as an indirect measure of cell density. (b) A spectrophotometer works by splitting white light from a source into a spectrum. The spectrophotometer allows choice of the wavelength of light to use for the measurement. The optical density (turbidity) of the sample will depend on the wavelength, so once one wavelength is chosen, it must be used consistently. The filtered light passes through the sample (or a control with only medium) and the light intensity is measured by a detector. The light passing into a suspension of bacteria is scattered by the cells in such a way that some fraction of it never reaches the detector. This scattering happens to a far lesser degree in the control tube with only the medium.

Measuring dry weight of a culture sample is another indirect method of evaluating culture density without directly measuring cell counts. The cell suspension used for weighing must be concentrated by filtration or centrifugation, washed, and then dried before the measurements are taken. The degree of drying must be standardized to account for residual water content. This method is especially useful for filamentous microorganisms, which are difficult to enumerate by direct or viable plate count.

As we have seen, methods to estimate viable cell numbers can be labor intensive and take time because cells must be grown. Recently, indirect ways of measuring live cells have been developed that are both fast and easy to implement. These methods measure cell activity by following the production of metabolic products or disappearance of reactants. Adenosine triphosphate (ATP) formation, biosynthesis of proteins and nucleic acids, and consumption of oxygen can all be monitored to estimate the number of cells.

Biofilms

Biofilm is an association of micro-organisms in which microbial cells adhere to each other on a living or non-living surfaces within a self-produced matrix of extracellular polymeric substance. Bacterial biofilm is infectious in nature and can results in nosocomial infections. According to National Institutes of Health (NIH) about about 65% of all microbial infections, and 80% of all chronic infections are associated with biofilms. Biofilm formation is a multi-step process starting with attachment to a surface then formation of micro-colony that leads to the formation of three dimensional structure and finally ending with maturation followed by detachment. During biofilm formation many species of bacteria are able to communicate with one another through specific mechanism called quorum sensing. It is a system of stimulus to co-ordinate different gene expression. Bacterial biofilm is less accessible to antibiotics and human immune system and thus poses a big threat to public health because of its involvement in variety of infectious diseases. A greater understanding of bacterial biofilm is required for the development of novel, effective control strategies thus resulting improvement in patient management.

Composition of Biofilms

Biofilms are group or micro-organisms in which microbes produced an extracellular polymeric substances (EPS) such as proteins (<1-2%) including enzymes), DNA (<1%), polysaccharides (1-2%) and RNA (<1%), and in addition to these components, water (up to 97%) is the major part of biofilm which is responsible for the flow of nutrients inside biofilm matrix. The architecture of biofilm consists of two main components i.e. water channel for nutrients transport and a region of densely packed cells having no prominent pores in it. The microbial cells with in biofilms are arranged in way with significant different physiology and physical properties. Bacterial biofilms are normally beyond the access of antibiotics and human immune system. Microorganisms that produce biofilm have enhanced potential to bear and neutralize antimicrobial agents and result in prolonged treatment. Biofilm forming bacteria switch on some genes that activate the expression of stress genes which in turn switch to resistant phenotypes due certain changes e.g. cell density, nutritional or temperature, cell density, pH and osmolarity. When the biofilm water channels are compared with system of circulations showed that biofilms are considered primitive multi-cellular organisms. Various components of biofilms are shown in table below signify the biofilm integrity and making it resistant against various environmental factors.

S. No	Components	Percentage of matrix
1	Microbial cells	2-5%
2	DNA and RNA	<1-2%
3	Polysaccharides	1-2%
4	Proteins	<1-2% (including enzymes)
5	Water	Up to 97%

Table: Biofilm chemical composition.

Formation of Biofilms

Biofilm formation is a highly complex process, in which microorganism cells transform from planktonic to sessile mode of growth. It has also been suggested that biofilm formation is dependent on the expression of specific genes that guide the establishment of biofilm. The process of biofilm formation occurs through a series of events leading to adaptation under diverse nutritional and environmental conditions. This is a multi-step process in which the microorganisms undergo certain changes after adhering to a surface. Microorganisms which form biofilms are shown to elicit specific mechanisms. Biofilm formation has following important steps (a) attachment initially to a surface (b) formation of micro-colony (c) three dimensional structure formation (d) biofilm formation, maturation and detachment (dispersal).

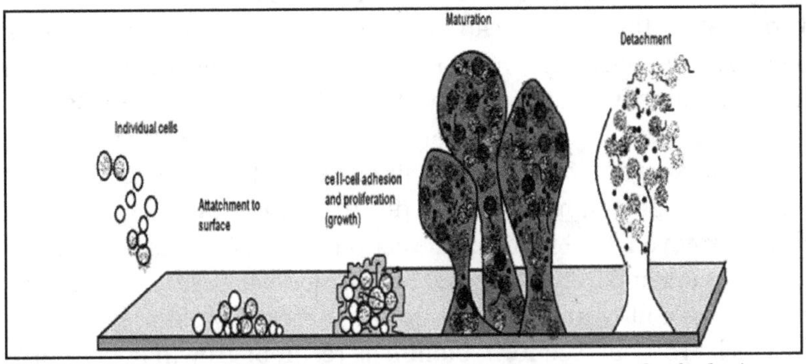

Figure: the biofilm life cycle in three steps: attachment, growth of colonies (micro-colony formation and formation of three dimensional structures) and detachment in clumps.

Attachment

When a bacterium cell reaches to near some surface/support so close that its motion is very slow down, it make a reversible connection with the surface and/or already adhered other microbe to the surface. For biofilm formation, a system of solid– liquid interface can provide an ideal environment for micro-organism to attach and grow (e.g. blood, water). For most frequent attachment and biofilm formation rough, hydrophilic and coated surfaces will provide better environment. Increased attachment may also occur due to increase but not exceeding critical level in flow velocity, temperature of water or nutrients concentrations. Presence of locomotor structures on cell surfaces such as flagella, pili, fimbriae, proteins or polysaccharides are also important and may possibly provide an advantage in biofilm formation when there are mixed community.

Micro-colony Formation

Micro-colony formation takes place after bacteria adhered to the physical surface/biological tissue and this binding then becomes stable which results in formation of micro-colony. Multiplication of bacteria in the biofilm starts as a result of chemical signals. The genetic mechanism of exopolysaccharide production is activated when intensity of the signal cross certain threshold. So by this way using such chemical signal, the bacterial cell divisions take place within the embedded exopolysaccharide matrix, which finally result in micro-colony formation.

Three-dimensional Structure Formation and Maturation

After micro-colony formation stage of biofilm, expression of certain biofilm related genes take place. These gene products are needed for the EPS which is the main structure material of biofilm. It is reported that bacterial attachment by itself can trigger formation of extracellular matrix.

Matrix formation is followed by water-filled channels formation for transport of nutrients within the biofilm. Researcher have proposed that these water channels are like a circulatory systems, distributing different nutrients to and removing waste materials from the communities in the micro-colonies of the biofilm.

Detachment

After biofilm formation, the researchers have often noticed that bacteria leave the biofilms itself on regular basis. By doing this the bacteria can undergo rapid multiplication and dispersal. Detachment of planktonic bacterial cells from the biofilm is a programmed detachment, having a natural pattern . Sometime occasionally due to some mechanical stress bacteria are detached from the colony into the surrounding. But in most cases some bacteria stop EPS production and are detached into environment. Dispersing of biofilm cells occur either by detachment of new formed cells from growing cells or dispersion of biofilm aggregates due to flowing effects or due to quorum- sensing . In biofilm of cells are removed due to an enzyme action that causes alginate digestion . Phenotypic characters of organisms are apparently affected by the mode of biofilm dispersion. Dispersed cells from the biofilm have the ability to retain certain properties of biofilm, such as antibiotic in-sensitivity. The cells which are dispersed form biofilm as result of growth may return quickly to their normal planktonic phenotype. The different steps in biofilm life cycle are shown in figure below.

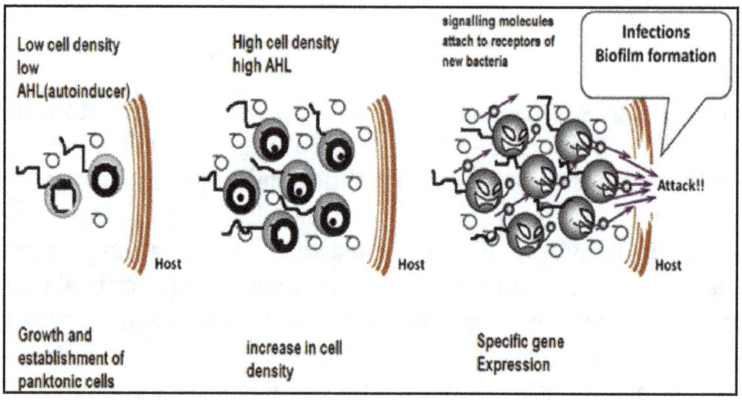

Figure: Cell density dependent gene expression in quorem sensing.

Quorum Sensing

During biofilm formation many species of bacteria are able to communicate with one another through a mechanism called quoreum sensing. It is a system of stimulus to co-ordinate gene expression with other cells and response related to the density of their local population. During quorum sensing signalling molecules attach to receptors of new bacteria and help in transcription of genes within a single species of bacteria as well as between different bacterial species. QS system enables communication between intraspecies and interspecies which involves in terms of biofilm formation, food shortages and environmental stress conditions, such as disinfectants, antibiotics, bacterial colonization, the identification of annoying species, the establishment of normal intestinal flora as well as the prevention of harmful intestinal flora. Many clinically-associated bacteria use QS for the regulation of the collective production of virulence factors. QS in Gram-positive bacteria occur through a series of events such as production, detection and response to AIs. The oligopeptide auto inducing peptides in many Gram-positive bacteria are detected by membrane-bound two component signal transduction systems. They are encoded as precursors (pro AIPs) and possess sequence diversity.

Biofilm Forming Bacteria

Nearly all (99.9%) of micro-organisms have the ability to form biofilm on a wide range of surfaces i.e. biological and inert surfaces. When micro-organisms bind to a surface, they produce extracellular polymeric substance (EPS) and form biofilm. Biofilm posing a great problem for public health due to its resistant nature to antibiotics and disease associated with indwelling medical devices. It is found that H. influenza has the ability to form biofilm in human body and can escape from human immune system . Biofilm forming capability has been reported in large number of bacterial species such as P. aeruginosa, S. epidermidis, E. coli spp, S. aureus, E. cloacae, K. pneumoniae. A few biofilm forming bacterial species have described below.

S.No	Common biofilm forming bacterial species
1	*E. coli*
2	*P. aeruginosa,*
3	*S. epidermidis,*
4	*S. aureus,*
5	*Staphylococcus epidermidis*
6	*E. cloacae*
7	*K. pneumoniae*
8	*Actenomyces israelii*
9	*Haemophilus influenza*
10	*Burkholderia cepacia*

Table: A list of common biofilm forming bacterial species.

Escherichia Coli

E. coli is a rod shaped Gram negative bacteria causing a large number of nosocomial and community infections such as urinary tract infections (UTIs) and prostatitis. It has the ability to secret toxins, polysaccharide and can form biofilm. It can also form biofilm in-vitro . E. coli capsules are high molecular weight molecules and are attached to the cell surface. E. coli capsule play an indirect role in biofilm by protecting bacterial surface adhesion. Different environmental conditions affect E. coli capability to form biofilm. Thickness of E. coli biofilm may be of hundreds of microns and posing a difficulty in treatment with antibiotics due to presence of exopolymers.

Pseudomonas Aeruginosa

P. aeruginosa is a Gram negative notorious opportunistic pathogen found along with other Pseudomonas species as part of normal flora of human skin. P. aeruginosa is a ubiquitous human pathogenic organism present everywhere, and can be isolated from different sources such as humans, plants and animals . It has a strong tendency to form biofilm and such biofilm has been found to be partially responsible for chronic infections. P. aeruginosa biofilm can be eradicated by using silver. Silver has antibacterial activity and bactericidal concentration of silver necessary to eradicate the bacterial biofilm was found to be 10-100 times higher as compared to that used to eradicate planktonic bacteria. The biofilms of P. aeruginosa are developed communities of individual cells that are encased in an extracellular polysaccharide matrix. These are extremely resistant to antibiotics. Their biofilm formation involves an initial attachment to a solid surface which leads to the formation of micro-colonies. These micro colonies differentiate into exopolysaccharide-encased, mature biofilms. P. aeruginosa is a multidrur resistantant bacteria even including ciprofloxacin, that are commonly used for the treatment of lung infections.

Staphylococcus Aureus

S. aureus is a multi-drug resistant bacteria causing a number of nosocomial infections. It grows on catheters and chronic wounds as biofilm. S. aureus recycles proteins for the formation of the extracellular matrix in the cytoplasm. The cytoplasmic proteins also working as matrix proteins allows enhanced flexibility and adaptation to S. aureus in forming biofilms in infectious conditions and could encourage the formation of mixed-species biofilms in chronic wounds.

Stereptococcus Epidermidis

S. epidermidis is well known as an opportunistic pathogen that has greater potential to cause infections in patients with immune-compromised state, intravenous drug abusers, AIDS patients, immuno-suppressive therapy patients and premature new born . During surgical implantation of polymeric devices the S.epidermidis contamination chances increases because of its biofilm forming capability. Biofilm formation is responsible for device related infections of S. epidermidis and this consequently leads to the pathogenesis. In staphylococci, the main factor which is responsible for adhesion is called the polysaccharide intercellular adhesion (PIA). In Staphylococcus cells are covers by PIA which hold them together as the most important component of the extracellular matrix. In a recent study Rohde have reported PIA-independent biofilm formation in about 27% of biofilm-forming S. epidermidis strains isolated from prosthetic joint infections.

Enterobacter Cloacae

E. cloacae is a Gram positive bacteria causing a range of nosocomial infections in human i.e. lower respiratory tract infection, bacteraemia, urinary tract infections, endocarditis, intra-abdominal infections, septic arthritis, skin and soft tissue infections, osteomyelitis and ophthalmic infections. Enterobacter causing nasocomial infections is most frequently isolated species and in recent years has emerged as important pathogenic bacteria. Bloodstream infections which are responsible for morbidity and mortality in both developing and developed countries are caused by E. cloacae. It also causes biofilm associated infections such as UTIs and biliary tract infections. Enterobector also has property of intrinsic resistance to certain antibiotics such as ampicillin and narrow-spectrum cephalosporins. It is also showing high frequency of mutations to expanded-spectrum and broadspectrum cephalosporin. It also possesses ß-lactamase and showing resistance to third generation cephalosporins.

Klebsiella Pneumoniae

K. pneumoniae is a Gram negative bacterium, frequently causing nosocomial infections, belongs to the genus Klebsiella. K. pneumonie is very important species among genus Klebsiella and causing a considerable proportion of nosocomial infections such as urinary tract infections (UTI), pneumonia, septicemias and soft tissue infections. Field emission scanning electron microscopy (F-ESEM) and confocal laser scanning microscopy were used to investigate the biofilm of K. pneumoniae, isolated from clinical strains. In a study on different strains of K. pneumonia isolated from various human samples such as urine, blood sputum and wound swabs, it was reported that about 40% strains had the ability to form biofilms. However, increasing temperature from 35°C to 40°C resulted in consistent growth of biofilm on abiotic surfaces. Also the ability of Klebsiella pneumonia to form biofilm takes place more successfully in a mix strains than individual strain.

Biofilms and Antibiotic Resistance

Mechanisms of antibiotics and biocides resistance of biofilms are categorized into four classes which include (a) active molecule inactivation directly (b) altering body's sensitivity to target of action, (c) reduction of the drug concentration before reaching to the target site and (d) efflux systems. Biofilm antibiotic resistance level may vary among different sittings and the key factors responsible for this resistance may also differ. Regarding resistance, the primary evidence shows that conventional mechanisms are unable to explain the high resistance to antibacterial agents associated with biofilms, although this evidence cannot be ignored in resistance in the growth of adherent cells. So it is suggested that the resistance posed by the adhered bacteria or biofilms may have some intrinsic mechanisms and are responsible for conventional antibiotic resistance. Several mechanisms have been explored that are considered to be key factors in high resistance nature of biofilms. These mechanisms are (a) limited diffusion, (b) enzyme causing neutralizations, (c) heterogeneous functions, (d) slow growth rate, (e) presence of persistent (non-dividing) cells and (f) biofilm phenotype such adaptive mechanisms e.g. efflux pump and membrane alteration.

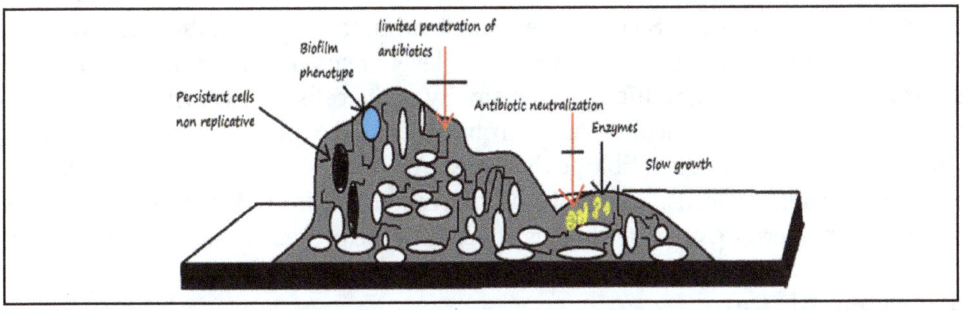

Figure: Antibiotic resistance associate to biofilm. Description of the key involved in antibiotic resistance such as enzyme causing neutralization, presence of persistent (non- dividing) cells and biofilm phenotype.

Low Penetration of Antibiotics

Diffusion of antibiotics can take place through the matrix of the biofilm. Diffusion or penetration of antibiotics to deeper layers of biofilm is affected by exopolysaccharide acting as a physical barrier. When molecules direct interact with this matrix, their movement to the interior of the biofilm is slow down, resulting antibiotic resistance. This may also acts as a hindrance for high molecular weight molecules such as complement system proteins and lysozyme, and in liquid culture bacterial cells are readily exposed to antibiotics as compare to compact structure biofilm. Bacteria escape from biofilm that do not produce polysaccharide and are easily attack by immune system cells.

Inactivation of antibiotic takes place when bind to biofilm matrix. P. aeruginosa have alginate exopolysaccharide, which is anionic in nature. Presence of this matrix explains slow penetration of fluoroquinolones and aminoglycosides. Low penetration of antibiotic is not sufficient to explain the biofilm resistance, other mechanisms have been assumed that must be involved. This is also suggested recently that slow diffusion of antibiotics permit plenty of time to establish a protective response to stress.

Neutralization by Enzymes

Antibiotics resistance in biofilm may be due to the presence of neutralizing enzymes which degrade or inactivate antibiotics. These enzymes are proteins which confer resistance by mechanisms such as hydrolysis, modification of antimicrobials by different biochemical reactions. Accumulations of these enzymes occur in the glycocalyx from the biofilm surface by the action of antibiotics. Neutralization by enzymes is enhanced by slow penetration of antibiotics and also antibiotics degradation in the biofilm. In cystic fibrosis which is caused by P. aeruginosa, overproduction of cephalosporinase AmpC enzymes is responsible for resistance to different antibiotics. This enzyme confers resistance to β-lactam in the presence of even much more concentration of carbapenems . During a study when filters impregnated with antibiotics was applied on K. pneumoniae biofilm (mutant cells β-lactamases), in spite of good diffusion of antibiotic, growth was observed, suggesting that there would be another mechanism of resistance which need to be explored.

Heterogeneous Nature

Studies performed on determination of microbial growth in biofilms by using a microelectrode with probes to direct measure oxygen concentration in different areas of the biofilms . The biofilms are heterogeneous nature both metabolically and structurally and both processes such as aerobic and anaerobic occur at the same time. So response against antibiotics may be different in different areas of the biofilms. On surface of biofilm there is a high level of activity of antibiotics while inside the biofilms, slow or absent growth reduces the sensitivity of the cells to antimicrobials. In the various sub layers of biofilm, aerobic or facultative anaerobic microbial populations help us to know the differential susceptibility to various antibiotic therapies. Antibiotics response to the planktonic forms is different from the adhered cells. Action of aminoglycosides is affected by limitation of oxygen and anaerobic growth of microorganisms, which is affected by the presence of oxygen and pH gradients.

Cells Slow Growth Rate

Slow growth of microorganisms occurs due to limited availability of nutrients which confer resistance to antibiotics. In case of biofilm a gradient of nutrients resulting in metabolically active cell (periphery or surface layer) and inactive cells (within its interior). Bacterial cells are attack by both penicillin and ampicillin only when they are growing. Some other antibiotics that attack cells in stationary phase are β-lactams, aminoglycosides, cephalosporin and fluoroquinolones . So due to slow growth resistance has been determined in different bacterial strains such as resistance to cetrimide on E. coli, piperacillin and tobramycin in P. aeruginosa and ciprofloxacin on S. epidermidis. It has been shown that this resistance was due to slow growth . There are some natural peptides produce during host innate immune response act as antibacterial providing protection to the body. A peptide named polymyxin E is effective in treating cancer patient's biofilms and cystic fibrosis, caused by P. aeruginosa . In Cystic fibrosis patients, using ciprofloxacin and tetracycline can clear active growing cells and it is suggested that a combination of antibiotic colistin with other two antibiotics (ciprofloxacin and tetracycline) will be very effective in clearing P. aeruginosa.

Existence of Persistent Cells

After a purling antibiotics treatment of biofilm, a very small number of bacterial cells remain viable, called persistent cells. These cells may or may not give this resistance to their progeny and return to their normal state after the release of the applied stress or pressure. The persistent cells stop their replication for small duration for the survival of the community. Their adaptive mechanism is not related to the mechanism followed by the cells during stress (environmental damage). Persistent cells can bear multiple antibiotic doses and work for survival. When density of bacterial cells number in stationary phase raised to maximum, persistent cells increase in number indicting their main role in survival. There are certain evidences for the presence of persistent cells in biofilm: a) there is existence of a biphasic dimension in biofilms which means that large number of cells population is attacked while the rest of the population is not attacked (resistant) even with an extensive antibiotics treatment, b) persistence gene description function as a circuits of regulation, c) bacteriostatic antibiotics contribute to the growth of persistent cell and biofilm preservation by inhibiting growth of sensitive cells and d) reshaping of biofilm into original form when the antibiotics therapy is withdrawn.

Biofilm Phenotype

During biofilm formation, bacteria produce some products called secondary metabolites. These products are not required by the cell for their growth. These metabolites function as signaling molecules thus enhancing formation process of biofilms. Biofilm phenotype is regarded as community cells that confer no response to antibiotics treatment. These characteristics have proposed the presence of specific genes. In B. subtilis only 6% difference in gene expression was observed for biofilm by DNA microarrays as compared to their planktonic culture cells, while for P. aeruginosa this difference was only 1%. However at present time, this differential gene expression has not been proven fruitful for describing this mechanism.

Efflux Pumps

Efflux pumps are protein structures, either express constitutively or intermittently. These pumps may have substrate specificity. Similar compounds can be transported by these pumps that may be involved in multidrug resistance. Efflux pumps, inside the bacteria in the periplasmic area, are involved in antagonized accumulation of antibiotics. The show resistant to multiple antibiotics such as tetracycline, macrolides, fluoroquinolones, β-lactam and thus reducing these antibiotics concentration at low toxic level. Five families of efflux transporters have been identified in prokaryotes. Over expression of the efflux pumps have been considered to be responsible for antibiotic resistance in P. aeruginosa biofilms.

Membrane Proteins Alterations

Permeability of outer membrane is very important for the antibiotic diffusion through different routes. A key role is played in transportation of hydrophilic molecules by outer membrane channel proteins (porins) present in Gram-negative bacteria from the outer environment to the periplasmic space. Mutation in porins encoding genes can result in production of non-functional or altered proteins. These mutant porins have low permeability for the passage of hydrophobic molecules . OprD is a specific porin present in P. aeruginosa which enhances the absorbance of basic amino acids and imipenem. Loss of OprD in P. aeruginosa is responsible for resistance to imipenem, resulting in three-dimensional disturbance of imipenem molecule. The differential expression of porins coding genes, occur in biofilm, leading to antibiotic resistance. When the expression of ompC and three other genes (osmotically regulated) is increased then the bacterial cells grow as biofilms.

Phase Variation

Diverse phenotype within biofilms plays a significant role and is responsible for the resistant infections. Authors have reported this phenomenon for many genera and species which are exemplified by Pseudomonas and Staphylococcus genus, and also some species of Enterobacteriaceae. Biofilms have the capability to develop bacterial subpopulation to switch to the quiescent state as small-colony known as small colony variants (SCVs). These variants cells have very less susceptibility to growth phase dependent killing of antibiotics. Also have a defective catalase activity which interferes with oxidative metabolism. Detectable colonial morphological changes caused by SCVs in biofilms, leads to increased adherence, auto-aggregation, increased hydrophobicity and low level motility. It can also withstand wide range of harsh environmental stress conditions so this is considered as survival mechanism for biofilms. Phase variation was considered as the process of cellular internal rearrangement but recently it is consider as phase variation that occur due to genetic elements interaction. Various mechanisms involved in antibiotic resistance due to biofilm.

Biofilm Role in Infections

Microscopic evaluation of specimens from chronic wounds often indicates the presence of biofilms. Traditionally, three phases are use to describe wound microbiology. These three phases described as: contamination, colonization, and infection. Contamination refers to the presence of bacteria in the wound, whereas the term colonization is used for bacterial community which is multiflying within the wound but not causing tissue damage. The “critical colonization” is used to describe bacteria that are growing inside the wound but do not possess classical symptoms of infection. However, they adversely affect wound healing. Some researchers have suggested that bacteria may play a vital role in normal wound healing. However, the specific role of bacterial community in wound healing is still under debate. The wound communities have polymicrobial nature. Bacteria can find its way to a wound from exogenous (soil and water) and endogenous (skin, saliva, urine, and faeces) sources. Bacteria that do multiply are not considered “infective” unless and until they pose detrimental effects. This is particularly the case for Corynebacteria and coagulase-negative Staphylococcus which are considered as commensal skin organisms.

Biofilm Formation in Acute Wound

Biofilm formation in wounds has been investigated in-vivo using model organisms (murine and porcine). Akiyama and co-workers investigated S. aureus and Streptococcus biofilm formation in mice wounds. These mice were treated with cyclophosphamide to inhibit leukocytes. Normal mice rapidly clear the inoculated bacteria because of a strong PMN response. P. aeruginosa biofilm, after examining burn wounds of murine model, shows that P. aeruginosa rapidly colonized burn wounds and formed biofilms around blood vessels.

Biofilm Formation in Chronic Wound

Bacterial biofilm interrupts the human immune system in several ways. In initial stages of chronic wound, antibiotic treatment is considered to be the immediate step. But in case of mature and established biofilm, antibiotic therapy is least effective and has only short term effects on both inflammation and healing. Clinicians have to depend on the results from a swab or biopsy, which barely represent all microorganisms present in the wound. The bacterial community residing in biofilm can be up to 1000 times resistant to antimicrobials. Even silver treatment which is considered to be quite effective now a days and incorporated in wound dressings, is least effective on biofilm.

Biofilm in Infectious Kidney Stones

About 15-20% kidney stones are responsible for urinary tract infection and these stones are formed as a result of interaction between bacteria and mineral substances derived from urine. This result in the formation of complex biofilm composed of infecting bacteria and their exoproducts, and mineralized stone material. Hellstrom in 1938 for the first time examined the stones that were collected from his patients and discovered the occurrence of bacteria deep inside them. Microscopic analysis of stones, which were removed from infected patients, have revealed that bacteria present in the stones are organized to form microcolonies surrounded by an anionic matrix composed of complex polysaccharides and minerals. Urine flow is obstructed by these infectious stones and thus causes severe inflammation and infection that can lead to kidney failure.

Bacterial Endocarditis

Bacteria and host components form complex biofilm that cause infection lesion in endocarditis. This biofilm is known as vegetation and can cause disease by three main mechanisms . First, the vegetation disrupts the function of valve by creating leakage, turbulence and flow of blood. Secondly, the vegetation causes bloodstream infections, and may lead to recurrent fever, chronic systemic inflammation, and other severe complications. Thirdly, sometimes vegetation breaks off into pieces and these pieces are then carried to extremities in the circulation system (embolization). Brain and kidneys are particularly vulnerable areas of the body. Vegetation is usually treated by prolonged administration of intravenous antibiotics or surgical excision of the infected valve.

Cystic Fibrosis Airway Infections

Cystic Fibrosis patients most commonly afflicted with P. aeruginosa airway infections. These infections are generally divided into two steps. First, intermittent respiratory infections are developed in CF patients. In second stage, permanent infections with P. aeruginosa take place which last for the rest of the patient's life, as confirmed by genetic fingerprinting. This persistant infection is clinically very important as it causes permanent failure of lungs.

Other Biofilm Diseases

Biofilms cause several diverse kinds of human infections. Biofilm is involved in otitis media with effusion. Fluid accumulates in the middle ear cavity of the patient, thus affecting speech development and learning capability of the patient. However, the complete etiology of the problem is still not clealy understood. In acute osteomyelitis, certain areas of bone necrose and produce favourable conditions for biofilm development. Biofilms have also been identified in most indwelling medical device infections and also in biliary tract infections, periodontitis, ophthalmic infections.

Immune Response to Biofilms

In-vitro studies have indicated that human leukocytes have the ability to penetrate biofilms of S. aureus. Akiyama reported that antimicrobial efficacy against S. aureus biofilms was comparatively more effective in studying acute wound infection in normal mice than those who had less leucocytes count. They investigated the primary mechanism of antimicrobials in normal mice, and found that it was due to the penetration of PMNs into the biofilm. In P. aeruginosa biofilms, production of the extracellular polysaccharide, alginate, protected them from IFN-γ mediated phagocytosis by human leukocytes, primarily known as monocytes. Similarly, polysaccharide intercellular adhesin (PIA) protected S. epidermidis against killing by polymorphonuclear leukocytes (PMN) and phagocytosis. Thus, extracellular polysaccharides play a significant role in biofilm resistance to phagocytosis. In addition, P. aeruginosa biofilms cause killing of PMN through the production of rhamnolipids. Thus limits the effectiveness of innate immune factors. Severe cutaneous infections are caused by S. aureus which is capable of producing leukotoxins, including the Panton-Valentine leukocidin.

Pilicides

Pili are extracellular fibers of bacterial cells which allow binding and colonization on epithelial cells. These pili are assembled via a specific mechanism called chaperone-usher pathway. Interupting this mechanism of pili assembly is a new and feasible approach due to its potential application in biofilm eradication. Researchers have also designed small synthetic compounds known as pilicides which inhibit the synthesis of pili.

Enzymes

Another effective way to degrade biofilm is the application of enzymes. Biofilm consists of extracellular polymeric substances (EPS, therefore these enzymes have the potential to degrade EPS. Biofilm is mainly composed of bacteria and EPS. When the biofilm is degraded by enzymes which results release of components and plaktonic cells which is easily clear by immune systems.

Inhibition of Quorum-sensing

Using inhibitors of quorum sensing is the most studied novel way in control of biofilm. In last decade, many researchers have searched for compounds that could block the QS system.

Electrical Currents

It has been shown that electrical current has antibacterial effect for several bacterial species. In a study by the application of low intensity current, a substantial reduction was observed in number of both viable bacteria of staphylococcus and pseudomonas biofilms. Electrical currents in combination with electromagnetic fields and ultrasound have given enhanced results on biofilm eradication in studies conducted in-vitro as well as in-vivo.

Surface Coatings

One of the most effective ways for eradication or blocking bacterial biofilm on surfaces of endotracheal tubes (ETTs) and catheters is coating the devices with metals, antiseptics or antimicrobials.

Bacteriophages

Use of bacteriophages for eradication/removal of biofilm is an efficient and novel strategy . Bacteriophages ability to inhibit or reduce formation of biofilm in-vivo has also been proven . During a study, when a genetically engineered lytic phage having a biofilm degrading enzyme, was used showed more efficient eradication of biofilm than wild type phages. Similarly a phage cocktail (combination of multiple phages) can also be used for efficient and complete eradication of bacterial biofilm.

Bacterial Conjugation

Bacterial conjugation is a way by which a bacterial cell transfers genetic material to another bacterial cell. The genetic material that is transferred through bacterial conjugation is a small plasmid, known as F-plasmid (F for fertility factor), that carries genetic information different from that which is already present in the chromosomes of the bacterial cell. In fact, the F-plasmid can replicate in the cytoplasm separately from the bacterial chromosome.

A cell that already has a copy of the F-plasmid is called an F-positive, F-plus or F^+ cell, and is considered a donor cell, while a cell that does not have a copy of the F-plasmid is called an F-negative, F-minus or F^- cell, and is considered a recipient cell. The transfer of the F-plasmid takes place through a horizontal connection by which the donor cell and the recipient cell directly contact each other or form a bridge between the two through which the genetic material is transferred. In cases where the F-plasmid of a donor cell has been integrated in the cell's genome (i.e., in the chromosome), a part of the chromosomal DNA may also be transferred to the recipient cell together with the F-plasmid.

Bacterial Conjugation Steps

In order to transfer the F-plasmid, a donor cell and a recipient cell must first establish contact. At this point, when the cells establish contact, the F-plasmid in the donor cell is a double-stranded DNA molecule that forms a circular structure. The following steps allow the transfer of the F-plasmid from one bacterial cell to another:

Step 1

The F^+ (donor) cell produces the pilus, which is a structure that projects out of the cell and begins contact with an F^- (recipient) cell.

Step 2

The pilus enables direct contact between the donor and the recipient cells.

Step 3

Because the F-plasmid consists of a double-stranded DNA molecule forming a circular structure, i.e., it is attached on both ends, an enzyme (relaxase, or relaxosome when it forms a complex with other proteins) nicks one of the two DNA strands of the F-plasmid and this strand (also called T-strand) is transferred to the recipient cell.

Step 4

In the last step, the donor cell and the recipient cell, both containing single-stranded DNA, replicate this DNA and thus end up forming a double-stranded F-plasmid identical to the original F-plasmid. Given that the F-plasmid contains information to synthesize pili and other proteins, the old recipient cell is now a donor cell with the F-plasmid and the ability to form pili, just as the original donor cell was. Now both cells are donors or F^+.

The four steps mentioned above can be seen in this figure:

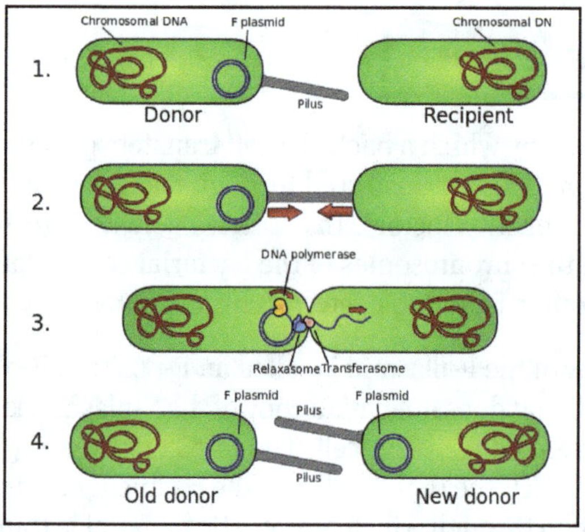

DNA Transfer

In order to avoid transferring the F-plasmid to an F^+ cell, the F-plasmid usually contains information that allows the donor cell to detect (and avoid) cells that already have one. In addition, the F-plasmid contains two main loci (tra and trb), an origin of replication (OriV) and an origin of transfer (OriT). The tra locus contains the genetic information to enable the donor cell to be attached to a recipient cell: the genes in the tra locus code for proteins to form the pili (pilin gene) in order to start the cell-cell contact, and other proteins to get attached to the F^- cell and to start the transfer of the F-plasmid. The trb locus contains DNA that codes for other proteins, such as some that are involved in creating a channel through which the DNA is transferred from the F^+ to the F^- cell. The OriV is the site at which replication of the DNA occurs and the OriT is the site at which the enzyme relaxase (or the relaxosome protein complex) nicks the DNA strand of the F-plasmid.

Although the DNA that is transferred in bacterial conjugation is that present in the F-plasmid, when the donor cell has integrated the F-plasmid into its own chromosomal DNA, bacterial conjugation can result in the transfer of the F-plasmid and of chromosomal DNA. When this is the case, a longer contact between the donor and the recipient cells results in a larger amount of chromosomal DNA being transferred.

The advantages of bacterial conjugation make this method of gene transfer a widely used technique in bioengineering. Some of the advantages include the ability to transfer relatively large sequences of DNA and not harming the host's cellular envelope. Furthermore, conjugation has been achieved in laboratories not only between bacteria, but also between bacteria and types of cells such as plant cells, mammalian cells and yeast.

References

- Bacterial-Genetics: biologyreference.com, Retrieved 02 March 2020
- Bacterial-growth: golifescience.com, Retrieved 24 May 2020
- How-microbes-grow, microbiology: courses.lumenlearning.com, Retrieved 10 July 2020
- Bacterial-conjugation: biologydictionary.net, Retrieved 04 May 2020
- Bacterial-biofilm-its-composition-formation-and-role-in-human-infections-61426: rroij.com, Retrieved 18 June 2020

Chapter 4

Bacterial Proteins

Bacteria contain a large number of proteins which are essential for the survival of the organism. They are arranged in network-like structures that perform distinct cellular functions. The various proteins found in bacteria like the AB toxin, cholera toxin, lac repressor, etc. are methodically elaborated in this chapter to help the reader gain a better perspective of the subject.

AB5 Toxin

The AB5 toxins are six-component protein complexes secreted by certain pathogenic bacteria known to cause human diseases such as cholera, dysentery, and hemolytic-uremic syndrome. One component is known as the A subunit, and the remaining five components make up the B subunit. All of these toxins share a similar structure and mechanism for entering targeted host cells. The B subunit is responsible for binding to receptors to open up a pathway for the A subunit to enter the cell. The A subunit is then able to use its catalytic machinery to take over the host cell's regular functions.

Families

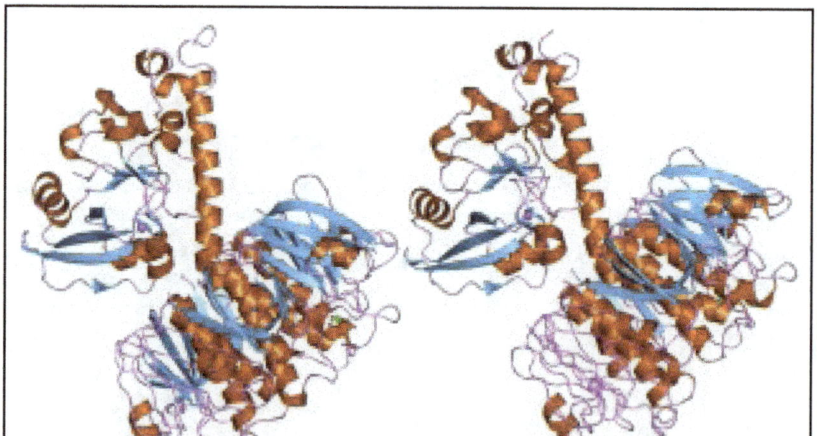

Ribbon diagram of cholera toxin. From PDB: 1s5e.

There are four main families of the AB5 toxin. These families are characterized by the sequence of their A subunit, as well as their catalytic ability.

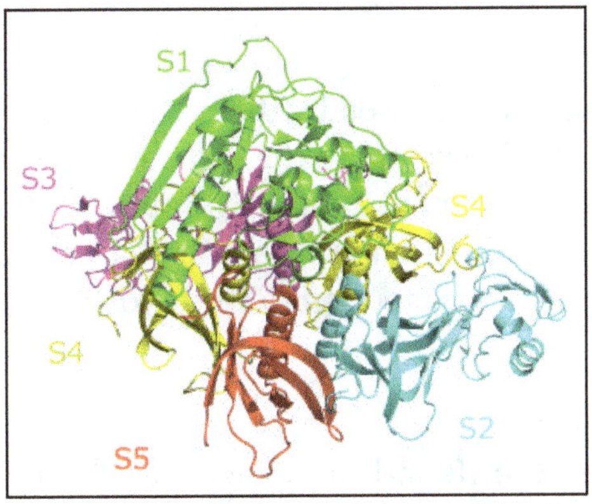

Ribbon diagram of pertussis toxin. S1 is the A subunit,and S2-S5 make up the B subunit.

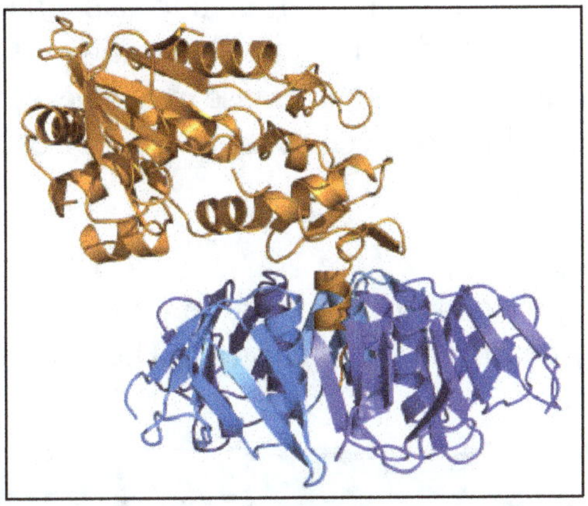

Ribbon diagram of shiga toxin (Stx) from *Shigella dysenteriae*, showing the characteristic AB5 structure. A subunit is in orange and B-subunit complex is in blue. From PDB: 1R4Q.

Cholera Toxin

This family is also known as Ct or Ctx, and includes the heat-labile enterotoxin family, known as LT. Cholera toxin's discovery is credited by many to Dr. Sambhu Nath De. He conducted his research in Calcutta (now Kolkata) making his discovery in 1959, although it was first purified by Robert Koch in 1883. Cholera toxin is an infectious toxin composed of a protein complex that is secreted by the bacterium *Vibrio cholerae*. Some symptoms of this toxin include chronic and widespread watery diarrhea and dehydration that, in some cases, leads to death.

Pertussis Toxin

This family is also known as Ptx and contains the toxin responsible for whooping cough. Pertussis toxin is secreted by the gram-negative bacterium, *Bordetella pertussis*. Whooping cough is very contagious and cases are slowly increasing in the United States despite vaccination. Symptoms include paroxysmal cough with whooping and even vomiting. The bacterium *Bordetella pertussis* was first identified as the cause of whooping cough and isolated by Jules Bordet and Octave Gengou in France in 1900.

Shiga Toxin

Shiga toxin is an infectious disease caused by the rod shaped *Shigella dysenteriae* as well as *Escherichia coli* (STEC), and is also known as Stx. Contaminated food and drinks are the source of infection and how this toxin is spread. Symptoms include abdominal pain as well as watery diarrhea. Severe life-threatening cases are characterized by hemorrhagic colitis (HC). The discovery of shiga toxin is credited to Dr. Kiyoshi Shiga in 1898.

Subtilase Cytotoxin

This family is also known as SubAB and was discovered during the 1990s. It produced by strains of STEC that do not have the locus of enterocyte effacement (LEE), and is known to cause hemolytic-uremic syndrome (HUS). It is called a subtilase cytotoxin because its A subunit sequence is similar to that of a subtilase-like serine protease in *Bacillus anthracis*. Some symptoms caused by this toxin are a decrease in platelet count in the blood or thrombocytopenia, an increase in white blood cell count or leukocytosis, and renal cell damage.

Structure

A complete AB5 toxin complex contains six protein units. Five units are similar or identical in structure and they comprise the B subunit. The last protein unit is unique and is known as the A subunit.

General diagram of the A subunit of the AB5 toxin with the disulfide linkage.

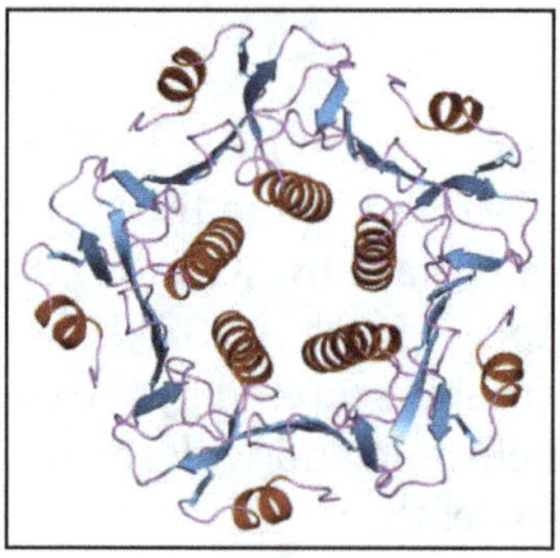
Ribbon diagram of the B-subunit of the cholera toxin.

A Subunit

The A subunit of an AB5 toxin is the portion responsible for catalysis of specific targets. For Shiga toxin family, the A subunit hosts a Trypsin-sensitive region which gives out two fragmented domains when cleaved. This region has not been confirmed for the other AB5 toxin families as yet. In general, the two domains of the A subunit, named A1 and A2, are linked by a disulfide bond. Domain A1 (approximately 22kDa in cholera toxin or heat labile enterotoxins) is the part of the toxin responsible for its toxic effects. Domain A2 (approximately 5kDa in cholera toxin or heat labile enterotoxin) provides a non-covalent linkage to the B subunit through the B subunit's central pore. The A1 chain for cholera toxin catalyzes the transfer of ADP-ribose from Nicotinamide adenine dinucleotide(NAD) to arginine or other guanidine compounds by utilizing ADP-ribosylation factors (ARFs). In the absence of arginine or simple guanidino compounds, the toxin mediated NAD+ nucleosidase (NADase) activity proceeds using water as a nucleophile.

B Subunit

The B subunits form a five-membered or pentameric ring, where one end of the A subunit goes into and is held. This B subunit ring is also capable of binding to a receptor on the surface of the host cell. Without the B subunits, the A subunit has no way of attaching to or entering the cell, and thus no way to exert its toxic effect. Cholera toxin, shiga toxin, and SubAB toxin all have B subunits that are made up of five identical protein components, meaning that their B subunits are homopentamers. Pertussis toxin is different where its pentameric ring is made up of four different protein components, where one of the components is repeated to form a heteropentamer.

Mechanisms

Cholera toxin, pertussis toxin, and shiga toxin all have their targets in the cytosol of the cell. After their B subunit binds to receptors on the cell surface, the toxin is enveloped by the cell and transported inside either through clathrin-dependent endocytosis or clathrin-independent endocytosis.

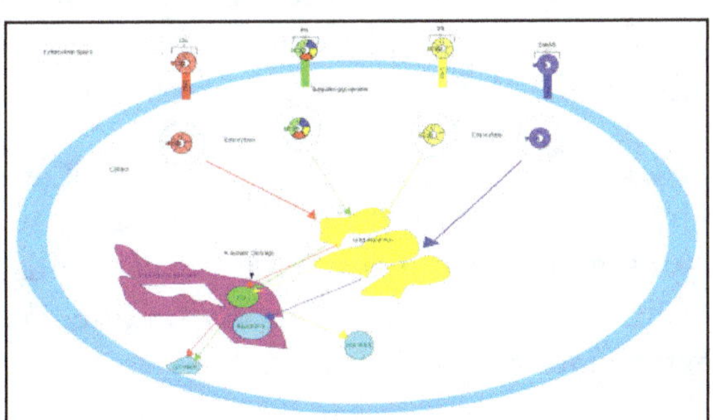

The mechanism pathways for the four AB5 toxins: cholera toxin, pertussis toxin, shiga toxin, and subtilase cytotoxin.

For the cholera toxin, the principal glycolipid receptor for the cholera toxin is ganglioside GM1. After endocytosis to the golgi apparatus, the toxin is redirected to the endoplasmic reticulum. In order for the A subunit to reach its target, a disulfide bond between the A1 and A2 domain must be broken. This breakage is catalyzed by a protein disulfide isomerase that is in the endoplasmic reticulum. Following separation, the A1 domain unfolds and is redirected back to the cytosol where it refolds and catalyzes ADP-ribosylation of certain G protein alpha subunits. In doing so, the downstream effects of the G protein signal transduction pathway is disrupted by activating adenylate cyclase. This causes a higher concentration of cAMP in the cell, which disrupts the regulation of ion transport mechanisms.

The pertussis toxin does not have a specific receptor, and binds to sialylated glycoproteins. After endocytosis, pertussis toxin's mechanism is the same as cholera toxin.

The main receptor for the shiga toxin is globotriaosylceramide or Gb3. Shiga toxin is also brought to the golgi apparatus before being directed to the endoplasmic reticulum for PDI to cleave the disulfide bond. Shiga toxin's A subunit is then brought back into the cytosol and inhibits eukaryotic protein synthesis with its RNA N-glycosidase activity by cleaving a specific adenine base on 28S ribosomal RNA that will ultimately cause cell death.

SubAB's target is in the endoplasmic reticulum of the cell and is brought into the cell through clathrin-mediated endocytosis. The glycan receptor for SubAB usually ends with an α2-3-linked N-Glycolylneuraminic acid (Neu5Gc). SubAB has an A subunit where it acts as a serine protease and cleaves Bip/GRP78, an endoplasmic reticulum chaperone. The cleavage of this chaperone causes cellular stress through protein inhibition, and consequently death of the cell.

Medical Uses

Cancer Treatment

B subunits of the AB5 toxins have the affinity towards binding glycan which some type of tumors seem to possess making it an easy target. One example is that of StxB which specifically binds with CD77 which shows expression on the surface of cancerous cells such as colon, pancreas, breast, and many more. Once StxB targets a cancerous cell, it delivers the A subunit of the toxin which eventually kills the cancerous cell.

Yet another method is by using ER stress-inducing drugs which have been tested in mice to show positive synergistic responses. This is accomplished through fusion of epidermal growth factor(EGF) with SubAB's A subunit. Cancer cells that express receptors for EGF will then experience SubAB toxicity.

Vaccines

Another use of AB5 toxins is using members of the LT family as adjuvants. This allows the toxin to promote immunological responses such as IgG2a, IgA, and Th17 to fight for instance gastric *Helicobacter pylori* infection when a vaccine is given.

In addition to some of these AB5 toxins being used to create vaccines to prevent bacterial infection, they are also being researched to work as a conjugate to prevent viral infections.

For example, systemic immunization along with co-administered intra-nasal delivery of virus-cholera toxin conjugate vaccine induced a virus-specific antibody response and showed some degree of protection to the upper respiratory tract from Sendai virus.

Recent Areas of Research

New advancements in biotechnological experimental methods such as the use of Bessel beam plane illumination microscopy and FRET-based sensor molecules can better demonstrate dynamic structures of gap junction plaques. For these experiments, different types of AB5 toxins can be used to induce the fast formation of tCDR in E.Coli cells. The response can then be recorded using cAMP concentration fluctuations in gap junction-coupled cells using FRET-based sensor constructs. Research suggests that CDRs could perhaps be linked with rapid rearrangement of lipids and protein in connexin channels within the gap junction plaques. This can further help us understand the signaling cascade that follows a cellular loss of K+ when exposed to bacterial infection.

The SubAB toxin has been seen to demonstrate specificity to a binding protein, BiP. This characteristic has been utilized to study the role of the cellular BiP itself, along with Endoplasmic-reticulum-associated degradation in stressed HeLa cells.

Actin Assembly-inducing Protein

The Actin assembly-inducing protein (ActA) is a protein encoded and used by *Listeria monocytogenes* to propel itself through a mammalian host cell. ActA is a bacterial surface protein comprising a membrane-spanning region. In a mammalian cell the bacterial ActA interacts with the Arp2/3 complex and actin monomers to induce actin polymerization on the bacterial surface generating an actin comet tail. The gene encoding ActA is named *actA* or prtB.

Introduction

As soon as *L. monocytogenes* bacteria are ingested by humans, they get internalized into intestinal epithelium cells and rapidly try to escape their internalization vacuole. In the cytosol they start to polymerize actin on their surface by the help of the ActA protein. It has been shown that ActA is not only necessary but also sufficient to induce motility of bacteria in the absence of other bacterial factors.

Discovery of ActA

ActA was discovered by analysing lecithinase-negative Tn*917-lac* Listeria mutants because of the phenotype that they were unable to spread from cell to cell. These mutant bacteria still escaped from the phagosomes as efficiently as wild-type bacteria and multiplied within the infected cells but they were not surrounded by actin like wild-type bacteria. Further analysis showed, that Tn*917-lac* had inserted into *actA*, the second gene of an operon. The third gene of this operon, *plcB*, encodes the *L. monocytogenes* lecithinase. To determine whether *actA* itself, *plcB* or other co-transcribed downstream regions are involved in actin assembly, mutations in the appropriate genes were generated. All mutants except the *actA* mutants were similar to wild-type concerning association with F-actin and cell-cell spreading. Complementation with *actA* restored wild-type phenotype in the *actA* mutants.

Function

ActA is a protein which acts as a mimic of Wiskott-Aldrich syndrome protein (WASP), a nucleation promoting factor (NPF) present in host cells. NPFs in the mammalian cell recruit and bind to the already existing actin-related-protein 2 and 3 complex (Arp2/3 complex) and induce an activating conformational change of the Arp2/3 complex. Due to this conformational change, NPFs initiate polymerization of a new actin filament at a 70° angle, which leads to the characteristic Y-branched actin structures in the leading edge of motile cells. ActA localizes to the old pole of the bacterium and spans both the bacterial cell membrane and the cell wall, lateral diffusion is inhibited; thus ActA localizes in a polarized and anchored manner on the bacterial surface. Consequently actin polymerization only starts in this region on the surface of the bacterium. Expression of ActA is induced only after entering a mammalian host cell.

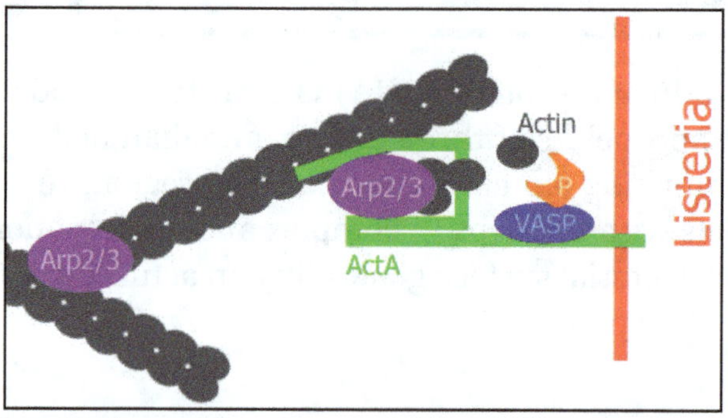

Fig. 1 Actin assembly induced by bacterial protein ActA (shown in green). Mammalian proteins involved in this process are: Profilin (P), Vasodilator-stimulated phosphoprotein (VASP) and actin-related-protein 2 and 3 complex (Arp2/3 complex) as well as actin.

Actin filament assembly generates the force that pushes the bacterium in the mammalian host cytoplasm forward. Continuous actin polymerization is sufficient for motility in the cytoplasm and even for infection of adjacent cells.

Research

New data indicates that ActA plays a role also in vacuolar disruption. A deletion mutant of ActA was defective in permeabilizing the vacuole. An 11 amino acid stretch of the N-terminus of the acidic region (32-42) was shown to be important for disruption of the phagosome.

Structure

The primary proteinous product of the *actA* gene consists of 639 amino acids and includes the signal peptide (first N-terminal 29 amino acids) and the ActA chain (C-terminal 610 amino acids). Therefore the sequence of the mature ActA protein consist of 610 amino acids. ActA has a molecular weight of 70,349 Da and is a surface protein.

ActA is a natively unfolded protein which can be divided into three functional domains:

- N-terminal domain that is highly charged: amino acid residues 1-234.
- central domain with proline-rich repeats: amino acid residues 235-394.
- C-terminal domain with a transmembrane domain: amino acid residues 395-610.

N-terminal Domain

The first 156 amino acids of the N-terminal domain consist of three regions:

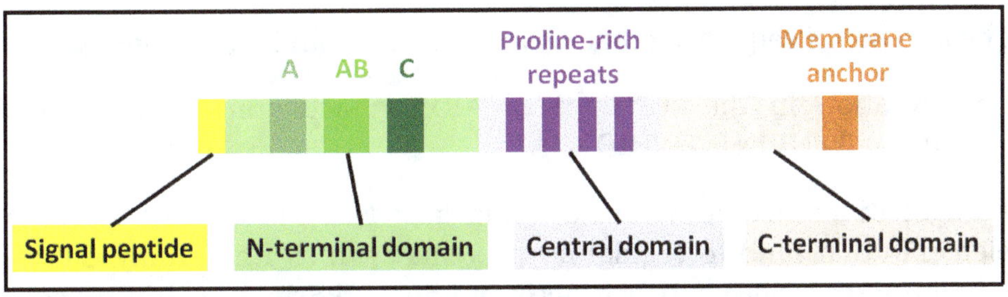

Fig. The ActA protein and its functional domains.

- A-region with a stretch of acidic residues: 32-45.
- AB-region, an actin monomer-binding region: 59-102.
- C-region, a cofilin homology sequence: 145-156.

The N-terminal portion of ActA plays an important role in actin polymerization. The domain displays consensus elements present in eukaryotic WASP family NPFs which include an actin monomer-binding region as well as an Arp2/3 binding C (central or cofilin homology) and A (acidic) region. The actin monomer-binding region of ActA has functional properties like the WASP-Homology-2 (WH2) or V domain, but differs in the sequence. Thus in WASP-family NPFs the order of the domains is WH2 followed by C,and then by A, which is not the case in ActA.

Central Domain

The central proline-rich region of ActA is crucial for ensuring efficient bacterial motility. There are four proline-rich repeats containing either FPPPP or FPPIP motifs. These regions mimic those of the host cell cytoskeletal protein zyxin, vinculin and palladin, known to associate with focal adhesions or stress fibers. The vasodilator-stimulated phosphoprotein (VASP) can bind through its Ena/VASP homology 1 domain (EVH1 domain) to the central proline-rich region and recruits profilin, an actin monomer binding protein, which itself promotes polymerization at barbed ends of actin filaments. Furthermore, VASP seems to interact with F-actin through its carboxy-terminal EVH2 domain, which provides a linkage of the bacterium to the tail. This statement is supported by the fact that ActA can bind multiple Ena/VASP proteins simultaneously and has a high affinity between ActA and Ena/VASP. VASP has been shown to reduce the frequency actin-Y-branches in vitro and thus increases the proportion of filaments which are organized in a parallel alignment in comet tails.

C-terminal Domain

The C-terminal domain of ActA has a hydrophobic region which anchors the protein in the bacterial membrane.

In summary, besides:

- The absence of sequence homology in the actin-binding-region.
- An alteration in the sequence of ARP2/3 activating domains typical for WASP-family NPFs (V(WH2)-C-A).
- A major difference between ActA and host NPFs is that ActA does not have elements that bind to regulatory proteins such as Rho family GTPases. This structural difference between ActA and host NPFs can be advantageous for *L. monocytogenes* and its pathogenesis because the actin nucleation activity of *L. monocytogenes* is independent of host regulation.

Analogues

WASP/N-WASP, which is functionally mimicked by ActA, is highly conserved in eukaryotes. It is an important actin-cytoskeleton organizer and is critical for processes such as endocytosis and cell motility. Activated by Cdc42, a Rho-family small GTPase, WASP/N-WASP activates the Arp2/3 complex, which leads to rapid actin polymerization.

Actin-based Motility of Other Pathogens

In *Shigella* the protein IcsA activates N-WASP, which in non-infected mammalian cells is activated by the GTPase Cdc42. Active N-WASP/WASP leads to actin polymerization by activating the Arp2/3 complex. In contrast, the *Listeria* ActA protein interacts with and activates directly the Arp2/3 complex.

The *Rickettsia* RickA protein is also able to activate the Arp2/3 complex in a WASP-like manner. In contrast to *Listeria*, the actin filaments are organized in long, unbranched parallel bundles. The Arp2/3 complex is only localized near the bacterial surface and thus it is assumed that a more frequent Arp2/3 complex-independent elongation occurs.

In *Burkholderia pseudomallei* BimA initiates actin polymerization in vitro. It is assumed that intracellular migration of this bacterium functions independently of the Arp2/3 complex.

Bacterial Effector Protein

Bacterial effectors are proteins secreted by pathogenic bacteria into the cells of their host, usually using a type 3 secretion system (TTSS/T3SS) or a type 4 secretion system (TFSS/T4SS). Some bacteria inject only a few effectors into their host's cells while others may inject dozens or even hundreds. Effector proteins may have many different activities, but usually help the pathogen to invade host tissue, suppress its immune system, or otherwise help the pathogen to survive. Effector proteins are usually critical for virulence. For instance, in the causative agent of plague (*Yersinia pestis*), the loss of the T3SS is sufficient to render the bacteria completely avirulent, even when they are directly introduced into the bloodstream. Gram negative microbes are also suspected to deploy bacterial outer membrane vesicles to translocate effector proteins and virulence factors via a novel membrane vesicle trafficking secretory pathway, in order to modify their environment or attack/invade target cells, for example, at the host-pathogen interface.

Diversity of Bacterial Effectors

Many pathogenic bacteria are known to have secreted effectors but for most species the exact number is unknown. Once a pathogen genome has been sequenced, effectors can be predicted based on protein sequence similarity, but such predictions are not always precise. More importantly, it is difficult to prove experimentally that a predicted effector is actually secreted into a host cell because the amount of each effector protein is tiny. For instance, Tobe et al. (2006) predicted more than 60 effectors for pathogenic *E. coli* but could only show for 39 that they are secreted into human Caco-2 cells. Finally, even within the same bacterial species, different strains often have different repertoires of effectors. For instance, the plant pathogen *Pseudomonas syringae* has 14 effectors in one strain, but more than 150 have been found in multiple different strains.

Species	number of effectors
Chlamydia (multiple species)	16+
E. coli EHEC (O157:H7)	40-60
E. coli (EPEC)	>20
Legionella pneumophila	>330 (T4SS)

Pseudomonas aeruginosa	4
Pseudomonas syringae	14 (>150 in multiple strains)
Salmonella enterica	60+
Yersinia (multiple species)	14

Mechanism of Action

Given the diversity of effectors, they affect a wide variety of intracellular processes. The T3SS effectors of pathogenic *E. coli, Shigella, Salmonella*, and *Yersinia* regulate actin dynamics to facilitate their own attachment or invasion, subvert endocytic trafficking, block phagocytosis, modulate apoptotic pathways, and manipulate innate immunity as well as host responses.

Phagocytosis. Phagocytes are immune cells that can recognize and "eat" bacteria. Phagocytes recognize bacteria directly [e.g., through the so-called scavenger receptor A which recognizes bacterial lipopolysaccharide (LPS)] or indirectly through antibodies (IgG) and complement proteins (C3bi) which coat the bacteria and are recognized by the Fcγ receptors and integrin$\alpha_m\beta_2$ (complement receptor 3). For instance, intracellular *Salmonella* and *Shigella* escape phagocytic killing through manipulation of endolysosomal trafficking. *Yersinia* predominantly survives extracellularly using the translocation of effectors to inhibit cytoskeletal rearrangements and thus phagocytosis. EPEC/EHEC inhibit both transcytosis through M cells and internalization by phagocytes. *Yersinia* inhibits phagocytosis through the concerted actions of several effector proteins, including YopE which acts as a RhoGAP and inhibits Rac-dependent actin polymerization.

Endocytic trafficking. Several bacteria, including *Salmonella* and *Shigella*, enter the cell and survive intracellularly by manipulating the endocytic pathway. Once internalized by host cells *Salmonella* subverts the endolysosome trafficking pathway to create a *Salmonella*-containing vacuole (SCV), which is essential for its intracellular survival. As the SCVs mature they travel to the microtubule organizing center (MTOC), a perinuclear region adjacent to the Golgi, where they produce *Salmonella* induced filaments (Sifs) dependent on the T3SS effectors SseF and SseG. By contrast, internalized *Shigella* avoids the endolysosome system by rapidly lysing its vacuole through the action of the T3SS effectors IpaB and C although the details of this process are poorly understood.

Secretory pathway. Some pathogens, such as EPEC/EHEC disrupt the secretory pathway. For instance, their effector EspG can reduce the secretion of interleukin-8 (IL-8), and thus affect the immune system (immunomodulation). EspG functions as a Rab GTPase-activating protein (Rab-GAP), trapping Rab-GTPases in their inactive GDP bound form, and reducing ER–Golgi transport (of IL-8 and other proteins).

Apoptosis (programmed cell death). Apoptosis is usually a mechanism of defense to infection, given that apoptotic cells eventually attract immune cells to remove them and the pathogen. Many pathogenic bacteria have developed mechanisms to prevent apoptosis, not the least to maintain their host environment. For instance, the EPEC/ EHEC effectors NleH and NleF block apoptosis. Similarly, the *Shigella* effectors IpgD and OspG (a homolog of NleH) block apoptosis, the former by phosphorylating and stabilizing the double minute 2 protein (MDM2) which in turn leads to a block of NF-kB-induced apoptosis. *Salmonella* inhibits apoptosis and activates pro-survival signals, dependent on the effectors AvrA and SopB, respectively.

Induction of cell death. In contrast to inhibition of apoptosis, several effectors appear to induce programmed cell death. For instance, EHEC effectors EspF, EspH, and Cif induce apoptosis.

Inflammatory response. Human cells have receptors that recognize pathogen-associated molecular patterns (PAMPs). When bacteria bind to these receptors, they activate signaling cascades such as the NF-kB and MAPK pathways. This leads to expression of cytokines, immunomodulating agents, such as interleukins and interferons which regulate immune response to infection and inflammation. Several bacterial effectors affect NF-kB signaling. For instance, the EPEC/EHEC effectors NleE, NleB, NleC, NleH, and Tir are immunosuppressing effectors that target proteins in the NF-kB signaling pathway. NleC has been shown to cleave the NF-kB p65 subunit (RelA), blocking the production of IL-8 following infection. NleH1, but not NleH2, blocks translocation of NF-kB into the nucleus. The Tir effector protein inhibits cytokine production. Similarly, YopE, YopP, and YopJ (in *Yersinia enterocolitica*, *Yersinia pestis*, and *Yersinia pseudotuberculosis* respectively) target the NF-kB pathway. YopE inhibits activation of NF-kB, which in part prevents the production of IL-8.

Databases and Online Resources

- Effectors.org – A database of predicted bacterial effectors. Includes an interactive server to predict effectors.
- Bacterial Effector Proteins and their domains/motifs (from Paul Dean's lab).
- T3DB – A database of Type 3 Secretion System (T3SS) proteins.
- T3SE – T3SS Database.
- BEAN 2.0: an integrated web resource for the identification and functional analysis of type III secreted effectors.

cAMP Receptor Protein

cAMP receptor protein (CRP; also known as catabolite activator protein, CAP) is a regulatory protein in bacteria. CRP protein binds cAMP, which causes a conformational change that allows CRP to bind tightly to a specific DNA site in the promoters of the genes it controls. CRP then activates transcription through direct protein–protein interactions with RNA polymerase.

The genes regulated by CRP are mostly involved in energy metabolism, such as galactose, citrate, or the PEP group translocation system. In *Escherichia coli*, cyclic AMP receptor protein (CRP) can regulate the transcription of more than 100 genes.

The signal to activate CRP is the binding of cyclic AMP. Binding of cAMP to CRP leads to a long-distance signal transduction from the N-terminal cAMP-binding domain to the C-terminal domain of the protein, which is responsible for interaction with specific sequences of DNA.

At "Class I" CRP-dependent promoters, CRP binds to a DNA site located upstream of core promoter elements and activates transcription through protein–protein interactions between "activating region 1" of CRP and the C-terminal domain of RNA polymerase alpha subunit. At "Class II" CRP-dependent promoters, CRP binds to a DNA site that overlaps the promoter -35 element and activates transcription through two sets of protein–protein interactions: (1) an interaction between "activating region 1" of CRP and the C-terminal domain of RNA polymerase alpha subunit, and (2) an interaction between "activating region 2" of CRP and the N-terminal domain of RNA polymerase alpha subunit. At "Class III" CRP-dependent promoters, CRP functions together with one or more "co-activator" proteins.

At most CRP-dependent promoters, CRP activates transcription primarily or exclusively through a "recruitment" mechanism, in which protein–protein interactions between CRP and RNA polymerase assist binding of RNA polymerase to the promoter.

Cholera Toxin

Cholera toxin (also known as choleragen and sometimes abbreviated to CTX, Ctx or CT) is protein complex secreted by the bacterium *Vibrio cholerae*. CTX is responsible for the massive, watery diarrhea characteristic of cholera infection.

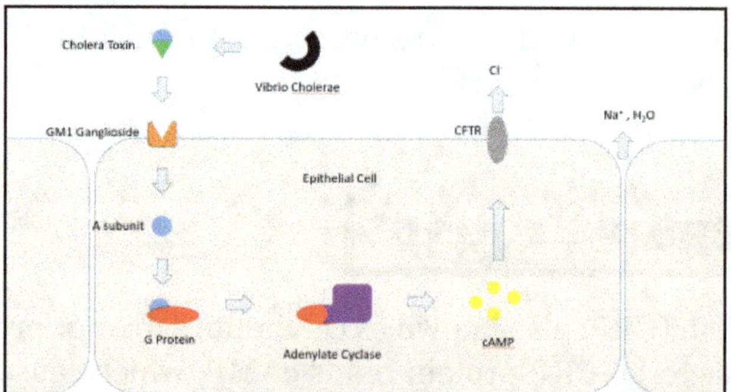

Cholera toxin mechanism.

History

Cholera toxin was discovered in 1959 by Late Prof.S.N.De (Ref.1) at Kolkata (India).

Structure

The cholera toxin is an oligomeric complex made up of six protein subunits: a single copy of the A subunit (part A, enzymatic), and five copies of the B subunit (part B, receptor binding), denoted as AB_5. Subunit B binds while subunit A activates the G protein which activates adenylate cyclase. The three-dimensional structure of the toxin was determined using X-ray crystallography by Zhang *et al.* in 1995.

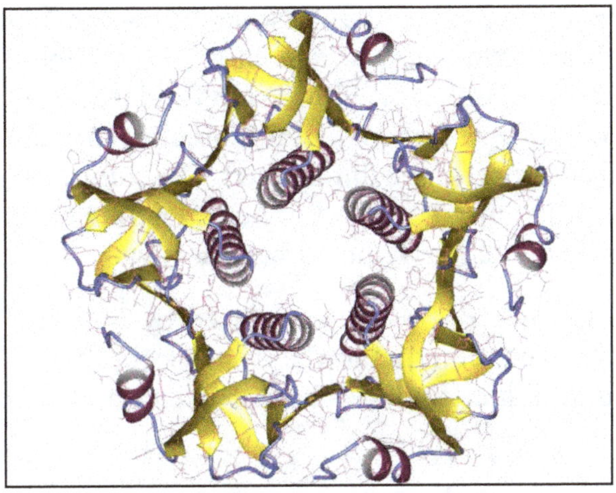

Cholera toxin B pentamer, Vibrio cholerae.

The five B subunits—each weighing 11 kDa, form a five-membered ring. The A subunit which is 28 kDa, has two important segments. The A1 portion of the chain (CTA1) is a globular enzyme payload that ADP-ribosylates G proteins, while the A2 chain (CTA2) forms an extended alpha helix which sits snugly in the central pore of the B subunit ring.

This structure is similar in shape, mechanism, and sequence to the heat-labile enterotoxin secreted by some strains of the *Escherichia coli* bacterium.

Pathogenesis

Cholera toxin acts by the following mechanism: First, the B subunit ring of the cholera toxin binds to GM1 gangliosides on the surface of target cells. The B subunit can also bind to cells lacking GM1. The toxin then most likely binds to other types of glycans attached to proteins instead of lipids. Once bound, the entire toxin complex is endocytosed by the cell and the cholera toxin A1 (CTA1) chain is released by the reduction of a disulfide bridge. The endosome is moved to the Golgi apparatus, where the A1 protein is recognized by the endoplasmic reticulum chaperon, protein disulfide isomerase. The A1 chain is then unfolded and delivered to the membrane, where Ero1 triggers the release of the A1 protein by oxidation of protein disulfide isomerase complex. As the

A1 protein moves from the ER into the cytoplasm by the Sec61 channel, it refolds and avoids deactivation as a result of ubiquitination.

CTA1 is then free to bind with a human partner protein called ADP-ribosylation factor 6 (Arf6); binding to Arf6 drives a change in the shape of CTA1 which exposes its active site and enables its catalytic activity. The CTA1 fragment catalyses ADP-ribosylation of the Gs alpha subunit ($G\alpha_s$) proteins using NAD. The ADP-ribosylation causes the $G\alpha_s$ subunit to lose its catalytic activity of inactivating GTP by hydrolyzing it to GDP + P_i, effectively increasing GTP concentration as less is being converted back to inactive GDP. Increased $G\alpha_s$ activation leads to increased adenylate cyclase activity, which increases the intracellular concentration of 3',5'-cyclic AMP (cAMP) to more than 100-fold over normal and over-activates cytosolic PKA. These active PKA then phosphorylate the cystic fibrosis transmembrane conductance regulator (CFTR) chloride channel proteins, which leads to ATP-mediated efflux of chloride ions and leads to secretion of H_2O, Na^+, K^+, and HCO_3^- into the intestinal lumen. In addition, the entry of Na^+ and consequently the entry of water into enterocytes are diminished. The combined effects result in rapid fluid loss from the intestine, up to 2 liters per hour, leading to severe dehydration and other factors associated with cholera, including a rice-water stool.

The pertussis toxin (also an AB_5 protein) produced by *Bordetella pertussis* acts in a similar manner with the exception that it ADP-ribosylates the $G\alpha_i$ subunit, rendering it inactive and unable to inhibit adenylyl cyclase production of cAMP (leading to constitutive production).

Expression of cholera toxin was shown to be prevented by the drug virstatin, which acts to prevent its transcription. It has been identified as a possible treatment for cholera.

Origin

The gene encoding the cholera toxin is introduced into *V. cholerae* by horizontal gene transfer. Virulent strains of *V. cholerae* carry a variant of temperate bacteriophage called the CTXf or CTXφ Bacteriophage.

Applications

Because the B subunit appears to be relatively non-toxic, researchers have found a number of applications for it in cell and molecular biology. It is routinely used as a neuronal tracer.

Treatment of cultured rodent neural stem cells with cholera toxin induces changes in the localization of the transcription factor Hes3 and increases their numbers.

GM1 gangliosides are found in lipid rafts on the cell surface. B subunit complexes labelled with fluorescent tags or subsequently targeted with antibodies can be used to identify rafts.

Crescentin

Crescentin is a protein which is a bacterial relative of the intermediate filaments found in eukaryotic cells. Just as tubulins and actins, the other major cytoskeletal proteins, have prokaryotic homologs in, respectively, the FtsZ and MreB proteins, intermediate filaments are linked to the crescentin protein.

Role in Cell Shape

Crescentin was recently discovered in the prokaryote *Caulobacter crescentus*, an aquatic bacterium which uses its crescent-shaped cells for enhanced motility. The crescentin protein is located on the concave face of these cells and appears to be necessary for their shape, as mutants lacking the protein form rod-shaped cells. To influence the shape of the *Caulobacter* cells, the helices of crescentin filaments associate with the cytoplasmic side of the cell membrane on one lateral side of the cell. This induces a curved cell shape in younger cells, which are shorter than the helical pitch of crescentin, but induces a spiral shape in older, longer cells.

Protein Structure

Like eukaryotic intermediate filaments, crescentin organizes into filaments and is present in a helical structure in the cell. Crescentin is necessary for both shapes of the *Caulobacter* prokaryote (vibroid/crescent-shape and helical shape, which it may adopt after a long stationary phase). The crescentin protein has 430 residues; its sequence mostly consists of a pattern of 7 repeated residues which form a coiled-coil structure. The DNA sequence of the protein has sections very similar to the eukaryotic keratin and lamin proteins, mostly involving the coiled-coil structure. Researchers Ausmees et al. recently proved that, like animal intermediate filament proteins, crescentin has a central rod made up of four coiled-coil segments. Both intermediate filament and crescentin proteins have a primary sequence including four α-helical segments along with non-α-helical linker domains. An important difference between crescentin and animal intermediate filament proteins is that crescentin lacks certain consensus sequence elements at the ends of the rod domain which are conserved in animal lamin and keratin proteins.

Assembly of Filaments

Eukaryotic intermediate filament proteins assemble into filaments of 8-15 nm within the cell without the need for energy input, that is, no need for ATP or GTP. Ausmees et al. continued their crescentin research by testing whether the protein could assemble into filaments in this manner *in vitro*. They found that crescentin proteins were indeed able to form filaments about 10 nm wide, and that some of these filaments organized laterally into bundles, just as eukaryotic intermediate filaments do. The similarity of crescentin protein to intermediate filament proteins suggests an evolutionary linkage between these two cytoskeletal proteins.

Iron-starvation-induced Protein A

IsiA is a photosynthesis-related chlorophyll-containing protein found in cyanobacteria. It belongs to the chlorophyll-a/b-binding family of proteins, and has been shown to have a photoprotection role in preventing oxidative damage via energy dissipation. It was originally identified under Fe-starvation, and thus received the name Iron-Starvation-Induced protein A. However, the protein has more recently been found to respond to a variety of stress conditions such as high irradiance. It can aggregate with carotenoids and form rings around the PSI reaction center complexes to aid in photoprotective energy dissipation.

Antenna Function

IsiA functions as an antenna for photosystem I (PSI) under iron-limiting conditions, when phycobilisomes disappear. In the (PSI)3(Isi3)18 complex most of the harvested energy is probably used by PSI; in other PSI-containing supercomplexes a large part of the energy will probably not be used for light harvesting, but rather is dissipated to protect the organism from light damage. Under iron-starvation, it forms a complex with PSI trimers, where the trimer is surrounded by a ring composed of 18 IsiA subunits. When the PSI subunits PsaF and PsaJ are missing the ring is composed of 17 IsiA subunits, indicating that each IsiA subunit has a different interaction with the trimer. This suggests that the size of the PSI complex determines the number of IsiA units in the surrounding ring. In the absence of PsaL, it has a tendency to form incomplete rings with PSI monomers, suggesting PsaL helps form the rings. Also it can form other aggregates of varying sizes depending on the level of iron-deprivation.

Photoprotective Function

IsiA aggregates, forming empty multimeric rings (without PSI) also accumulate and are very abundant in long-term iron-depleted cells. When isolated, these aggregates are in a strongly quenched state, suggesting they are responsible for thermal dissipation of absorbed energy. IsiA is also synthesized in cells grown under high light irradiances, protecting from photodamage.

Lac Repressor

The *lac* repressor is a DNA-binding protein which inhibits the expression of genes coding for proteins involved in the metabolism of lactose in bacteria. These genes are repressed when lactose is not available to the cell, ensuring that the bacterium only invests energy in the production of machinery necessary for uptake and utilization of lactose when lactose is present. When lactose becomes available, it is converted into allolactose, which inhibits the *lac* repressor's DNA binding ability. Loss of DNA binding by the *lac* repressor is required for transcriptional activation of the operon.

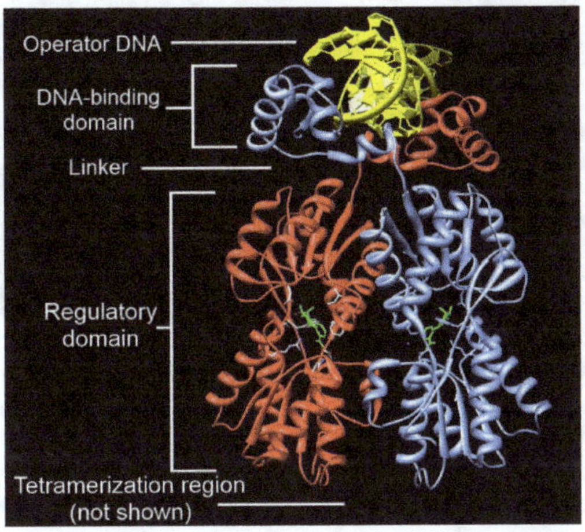

Annotated crystal structure of dimeric LacI. Two monomers (of four total) co-operate to bind each DNA operator sequence. Monomers (red and blue) contain DNA binding and core domains (labeled) which are connected by a linker (labeled). The C-terminal tetramerization helix is not shown. The repressor is shown in complex with operator DNA (gold) and ONPF (green), an anti-inducer ligand (*i.e.* a stabilizer of DNA binding).

Function

The *lac* repressor (LacI) operates by a helix-turn-helix motif in its DNA binding domain binding base-specifically to the major groove of the operator region of the *lac* operon, with base contacts also made by residues of symmetry-related alpha helices, the "hinge" helices, which bind deeply in the minor groove. This DNA binding causes the specific affinity of RNA polymerase for the promoter sequence to increase sufficiently that it cannot escape the promoter region and enter elongation, and so prevents transcription of the mRNA coding for the Lac proteins. When lactose is present, allolactose binds to the *lac* repressor, causing an allosteric change in its shape. In its changed state, the *lac* repressor is unable to bind tightly to its cognate operator. This effect is referred to as induction, because it *induces*, rather than represses, expression of the metabolic genes. *In vitro,* Isopropyl β-D-1-thiogalactopyranoside (IPTG) is a commonly used allolactose mimic which can be used to induce transcription of genes being regulated by *lac* repressor.

Structure

Structurally, the *lac* repressor protein is a homo-tetramer. The tetramer contains two DNA binding subunits composed of two monomers each (sometimes called "dimeric *lac* repressor"). These subunits dimerize to form a tetramer capable of binding two operator sequences.

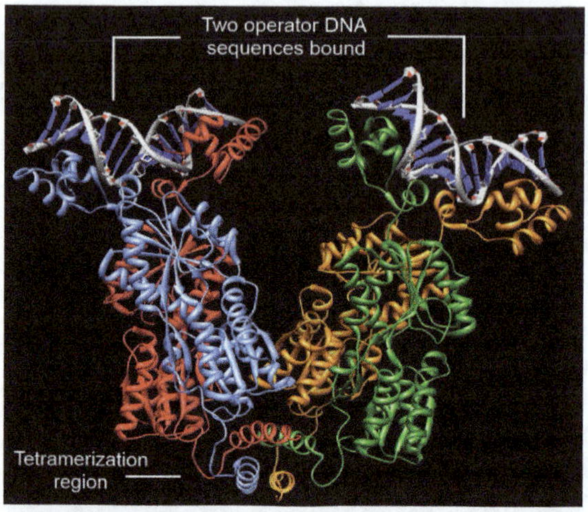

Tetrameric LacI binds two operator sequences and induces DNA looping. Two dimeric *LacI* functional subunits (red+blue and green+orange) each bind a DNA operator sequence (labeled). These two functional subunits are coupled at the tetramerization region (labeled); thus, tetrameric *LacI* binds two operator sequences. This allows tetrameric *LacI* to induce DNA looping.

Each monomer consists of four distinct regions:

- An N-terminal DNA-binding domain (in which two LacI proteins bind a single operator site).
- A regulatory domain (sometimes called the core domain, which binds allolactose, an allosteric effector molecule).
- A linker that connects the DNA-binding domain with the core domain (sometimes called the hinge helix, which is important for allosteric communication).
- A C-terminal tetramerization region (which joins four monomers in an alpha-helix bundle).

DNA binding occurs via an N-terminal helix-turn-helix structural motif and is targeted to one of several operator DNA sequences (known as O_1, O_2 and O_3). The O_1 operator sequence slightly overlaps with the promoter, which increases the affinity of RNA polymerase for the promoter sequence such that it cannot enter elongation and remains in Abortive initiation. Additionally, because each tetramer contains two DNA-binding subunits, binding of multiple operator sequences by a single tetramer induces DNA looping.

Discovery

The *lac* repressor was first isolated by Walter Gilbert and Benno Müller-Hill in 1966. They were able to show, *in vitro*, that the protein bound to DNA containing the *lac* operon, and released the DNA when IPTG was added. (IPTG is an allolactose analog.) They were also able to isolate the portion of DNA bound by the protein by using the enzyme deoxyribonuclease, which breaks down DNA. After treatment of the repressor-DNA complex, some DNA remained, suggesting that it had been masked by the repressor. This was later confirmed.

These experiments confirmed the mechanism of the *lac* operon, earlier proposed by Jacques Monod and Francois Jacob.

Methyl-accepting Chemotaxis Protein

Methyl-accepting chemotaxis protein (MCP) is a transmembrane sensor protein of bacteria. Use of the MCP allows bacteria to detect concentrations of molecules in the extracellular matrix so that the bacteria may smooth swim or tumble accordingly. If the bacteria detects rising levels of attractants (nutrients) or declining levels of repellents (toxins), the bacteria will continue swimming forward, or smooth swimming. If the bacteria detects declining levels of attractants or rising levels of repellents, the bacteria will tumble and re-orient itself in a new direction. In this manner, a bacterium may swim towards nutrients and away from toxins.

Environmental diversity gives rise to diversity in bacterial signalling receptors, and consequently there are many genes encoding MCPs. For example, there are four well-characterised MCPs found in *Escherichia coli*: Tar (taxis towards aspartate and maltose, away from nickel and cobalt), Tsr (taxis towards serine, away from leucine, indole and weak acids), Trg (taxis towards galactose and ribose) and Tap (taxis towards dipeptides).

Structure

MCPs share similar structure and signalling mechanism. MCPs form dimers. Three dimers of MCP spontaneously form trimers. Trimers are complexed by CheA and CheW into hexagonal lattices. MCPs either bind ligands directly or interact with ligand-binding proteins, transducing the signal to downstream signalling proteins in the cytoplasm. Most MCPs contain: (a) an N-terminal signal peptide that is a transmembrane alpha-helix in the mature protein; (b) a poorly-conserved periplasmic receptor (ligand-binding) domain; (c) a transmembrane alpha-helix; (d) generally one or more HAMP domains and (e) a highly conserved C-terminal cytoplasmic domain that interacts with downstream signalling components. The C-terminal domain contains the methylated glutamate residues.

MCPs undergo two covalent modifications: deamidation and reversible methylation at a number of glutamate residues. Attractants increase the level of methylation, while repellents decrease it. The methyl groups are added by the methyl-transferase CheR and are removed by the methylesterase CheB.

Function

Binding a ligand causes a conformational change in the MCP receptor which translates down the hairpin structure. At the tip of the hairpin are two proteins that associate to the MCP: CheW and CheA. CheA acts as the sensor kinase. CheA has kinase activity and autophosphorylates itself on a histidyl residue when activated by the MCP. CheW is believed to be a transducer of the signal from the MCP to CheA. Activated CheA transfers its phosphoryl group to CheY, a response regulator. Phosphorylated CheY phosphorylates the basal body FliM which is connected to the flagellum. Phosphorylation of the basal body acts as a flagellar switch and changes the direction of rotation of the flagellum. This change in direction allows for alternation between smooth swimming and tumbling which biases the bacterial random walk towards attractant.

Chapter 5

Diseases Caused by Bacteria

While most bacteria are useful, some cause various illnesses and diseases in other organisms. Bacterial diseases occur when these harmful bacteria get into a body and crowd out the healthy ones. Some bacterial diseases are bacterial pneumonia, actinomycosis, leprosy, tuberculosis, etc. This chapter discusses these bacterial diseases and their causes in detail.

Bacterial Pneumonia

Bacterial pneumonia is a type of pneumonia caused by bacterial infection.

Signs and Symptoms

- Pneumonia.
- Fever.
- Rigors.
- Cough.
- Runny nose (either direct bacterial pneumonia or accompanied by primary viral pneumonia).
- Dyspnea - shortness of breath.
- Chest pain.
- Shaking chills.
- Pneumococcal pneumonia can cause coughing up of blood, or hemoptysis, characteristically associated with "rusty" sputum.

Types

Gram-positive

Streptococcus pneumoniae (J13) is the most common bacterial cause of pneumonia in all age groups except newborn infants. *Streptococcus pneumoniae* is a Gram-positive bacterium that often lives in the throat of people who do not have pneumonia.

Other important Gram-positive causes of pneumonia are *Staphylococcus aureus* (J15.2) and *Bacillus anthracis.*

Gram-negative

Gram-negative bacteria are seen less frequently: *Haemophilus influenzae* (J14), *Klebsiella pneumoniae* (J15.0), *Escherichia coli* (J15.5), *Pseudomonas aeruginosa* (J15.1), *Bordetella pertussis*, and *Moraxella catarrhalis* are the most common.

These bacteria often live in the gut and enter the lungs when contents of the gut (such as vomit or faeces) are inhaled.

Atypical

"Atypical" bacteria are *Coxiella burnetii, Chlamydophila pneumoniae* (J16.0), *Mycoplasma pneumoniae* (J15.7), and *Legionella pneumophila.*

Many people falsely believe they are called "atypical" because they are uncommon and/or do not respond to common antibiotics and/or cause atypical symptoms. In reality, they are "atypical" because they do not gram stain as well as gram-negative and gram-positive organisms.

Pneumonia caused by *Yersinia pestis* is usually called pneumonic plague.

Pathophysiology

Bacteria typically enter the lung with inhalation, though they can reach the lung through the bloodstream if other parts of the body are infected. Often, bacteria live in parts of the upper respiratory tract and are continuously being inhaled into the alveoli. Once inside the alveoli, bacteria travel into the spaces between the cells and also between adjacent alveoli through connecting pores. This invasion triggers the immune system to respond by sending white blood cells responsible for attacking microorganisms (neutrophils) to the lungs. The neutrophils engulf and kill the offending organisms but also release cytokines that result in a general activation of the immune system. This results in the fever, chills, and fatigue common in bacterial and fungal pneumonia. The neutrophils, bacteria, and fluid leaked from surrounding blood vessels fill the alveoli and result in impaired oxygen transportation.

Bacteria often travel from the lung into the blood stream (bacteremia) and can result in serious illness such as sepsis and eventually septic shock, in which there is low blood pressure leading to damage in multiple parts of the body including the brain, kidney, and heart. They can also travel to the area between the lungs and the chest wall, called the pleural cavity.

Treatment

Antibiotics are the treatment of choice for bacterial pneumonia and ventilation (oxygen supplement) as supportive therapy. The antibiotic choice depends on the nature of the pneumonia, the microorganisms most commonly causing pneumonia in the geographical region, and the immune status and underlying health of the individual. In the United Kingdom, amoxicillin is used as first-line therapy in the vast majority of patients acquiring pneumonia in the community, sometimes with added clarithromycin. In North America, where the "atypical" forms of community-acquired pneumonia are becoming more common, clarithromycin, azithromycin, or fluoroquinolones as single therapy, have displaced the amoxicillin as first-line therapy. Local patterns of antibiotic-resistance should always be considered when initiating pharmacotherapy. In hospitalized individuals or those with immune deficiencies, local guidelines determine the selection of antibiotics.

Gram-positive Organisms

Streptococcus pneumoniae — amoxicillin (or erythromycin in patients allergic to penicillin); cefuroxime and erythromycin in severe cases. *Staphylococcus aureus* — flucloxacillin (to counteract the organism's β-lactamase).

Gram-negative Organisms

- *Haemophilus influenzae* — doxycycline; 2nd generation Cephalosporins such as Cefaclor.
- *Klebsiella pneumoniae.*
- *Escherichia coli.*
- *Pseudomonas aeruginosa* — ciprofloxacin.
- *Moraxella catarrhalis.*

Atypical Organisms

- *Chlamydophila pneumoniae* — doxycycline.
- *Chlamydophila psittaci* — erythromycin.
- *Mycoplasma pneumoniae* — erythromycin.
- *Coxiella burnetti* — doxycycline.
- *Legionella pneumophila* — erythromycin, with rifampicin sometimes added.

People who have difficulty breathing due to pneumonia may require extra oxygen. An extremely sick individual may require artificial ventilation and intensive care as life-saving measures while his or her immune system fights off the infectious cause with the help of antibiotics and other drugs.

Prevention

Prevention of bacterial pneumonia is by vaccination against streptococcus pneumoniae (pneumococcal polysaccharide vaccine for adults and pneumococcal conjugate vaccine for children), Haemophilus influenzae type B, meningococcus, bordetella pertussis, bacillus anthracis, and yersinia pestis.

Ehrlichia Ruminantium

Heartwater (also known as cowdriosis, nintas and ehrlichiosis) is a tick-borne rickettsial disease of domestic and wild ruminants. It is caused by *Ehrlichia ruminantium* (formerly *Cowdria ruminantium*) - an intracellular gram-negative coccal bacterium (also referred to as *Rickettsia ruminantium*). The disease is spread by bont ticks, which are members of the genus *Amblyomma*. Affected mammals include cattle, sheep, goats, antelope, and buffalo, but the disease has the biggest economic impact on cattle production in affected areas. The disease's name is derived from the fact that fluid can collect around the heart or in the lungs of infected animals.

The disease is common in sub-Saharan Africa and some of the West Indian islands. It was first identified in sheep in South Africa in the 1830s, and had reached the Caribbean by 1980. The ticks which carry the disease occur in Africa and the Caribbean, and feed on a wide variety of vertebrate hosts. In the Caribbean, at least, the cattle egret has been implicated in the spread of heartwater since it colonized the islands in the 1950s. Animals often acquire the disease when moved on to heartwater infected grazing.

Cowdriosis is notifiable to the World Organisation for Animal Health.

Clinical Signs

Clinical disease is more common in young animals and non-native breeds. The clinical signs of disease are caused by an increased vascular permeability and consequent oedema and hypovolaemia.

The symptoms include neurological signs such as tremors and head pressing, respiratory signs such as coughing and nasal discharge, and systemic signs such as fever and loss of appetite. Physical examination may reveal petechiae of the mucous membranes, tachycardia and muffled heart sounds. Cowdriosis can also cause reproductive and gastrointestinal disease. It is frequently fatal.

Diagnosis

On post-mortem examination, a light yellow transudate that coagulates on exposure to air is often found within the thorax, pericardium and abdomen. Most fatal cases will have the hydropericardium that gives the disease its common name. Pulmonary oedema and mucosal congestion are regularly seen along with frothy fluid in the airways and cut surfaces of the lungs.

To definitively diagnose the disease, *C. ruminantium* must be demonstrated either in preparations of the hippocampus under Giemsa staining or by histopathology of brain or kidney.

Treatment and Control

During the early stages of disease animals may be treated with sulfonamides and tetracyclines. In advanced disease, prognosis is poor.

Amblyomma hebraeum, a vector of Heartwater disease.

Tetracyclines can also be used prophylactically when animals are introduced into an area endemic with cowdriosis. A live blood vaccine is available for protection of young stock, but animals may require treatment for the disease post-vaccination. Ectoparasiticides dips can be used to reduce exposure the animals exposure to bont ticks. In areas endemic for heartwater there is likely to be use of dips against other ticks of domestic animals, such as *Rhipicephalus (Boophilus)* and *Hyalomma* species and this will usually contribute to control of vectors of *Ehrlichia ruminantium*.

Actinomycosis

Actinomycosis is a rare infectious bacterial disease caused by *Actinomyces* species. About 70% of infections are due to either *Actinomyces israelii* or *A. gerencseriae*. Infection can also be caused by other Actinomyces species, as well as *Propionibacterium propionicus*, which presents similar symptoms. The condition is likely to be polymicrobial aerobic anaerobic infection.

Signs and Symptoms

The disease is characterised by the formation of painful abscesses in the mouth, lungs, or gastrointestinal tract. Actinomycosis abscesses grow larger as the disease progresses, often over months. In severe cases, they may penetrate the surrounding bone and muscle to the skin, where they break open and leak large amounts of pus, which often contains characteristic granules (sulfur granules) filled with progeny bacteria. These granules are named due to their appearance but are not actually composed of sulfur. Sometimes there is difficulty in making the correct diagnosis. In addition to microbiological examinations, magnetic resonance imaging and immunoassays may also be helpful.

Causes

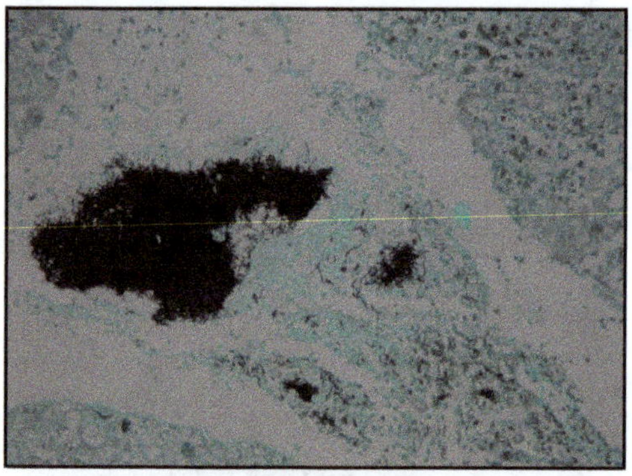

Actinomycosis Grocott's stain.

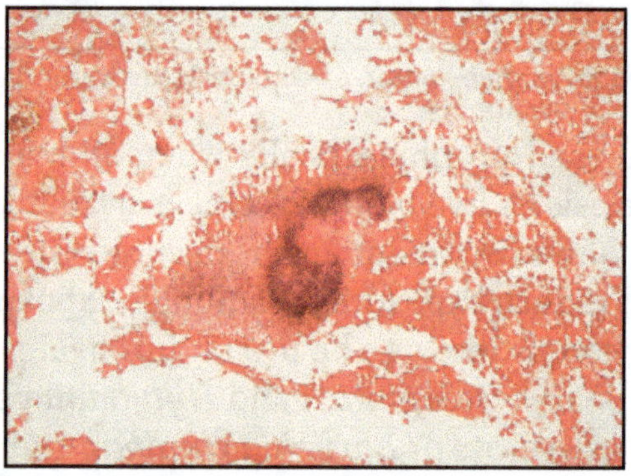

Actinomycosis Gram stain.

Actinomycosis is primarily caused by any of several members of the bacterial genus *Actinomyces*. These bacteria are generally anaerobes. In animals, they normally live in the small spaces between the teeth and gums, causing infection only when they can multiply freely in anoxic environments. An affected human often has recently had dental work, poor oral hygiene, periodontal disease, radiation therapy, or trauma (broken jaw) causing local tissue damage to the oral mucosa, all of which predispose the person to developing actinomycosis. *Actinomyces israelii* is a normal commensal species part of the microbiota species of the lower reproductive tract of women. They are also normal commensals among the gut flora of the caecum; thus, abdominal actinomycosis can occur following removal of the appendix. The three most common sites of infection are decayed teeth, the lungs, and the intestines. It is important to note that actinomycosis does not occur in isolation from other bacteria. This infection depends on other bacteria (gram positive, gram negative, and cocci) to aid in invasion of tissue.

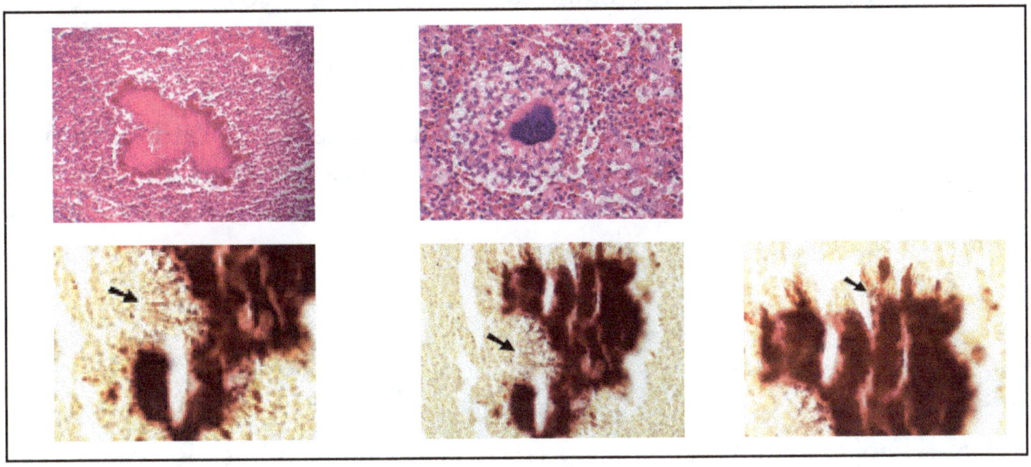

Treatment

Actinomyces bacteria are generally sensitive to penicillin, which is frequently used to treat actinomycosis. In cases of penicillin allergy, doxycyclin is used. Sulfonamides such as sulfamethoxazole may be used as an alternative regimen at a total daily dosage of 2-4 grams. Response to therapy is slow and may take months. Hyperbaric oxygen therapy may also be used as an adjunct to conventional therapy when the disease process is refractory to antibiotics and surgical treatment.

Epidemiology

There is a greater disease incidence in males between the ages of 20 and 60 years than in females. Before antibiotic treatments became available, the incidence in the Netherlands and Germany was 1 per 100,000 people/year. Incidence in the U.S. in the 1970s was 1 per 300,000 people/year, while in Germany in 1984, it was estimated to be 1 per 40,000 people/year. The use of intrauterine devices (IUDs) has increased incidence of genitourinary actinomycosis in females. Incidence of oral actinomycosis, which is harder to diagnose, has increased.

History

In 1877, pathologist Otto Bollinger described the presence of *Actinomyces bovis* in cattle, and shortly afterwards, James Israel discovered *Actinomyces israelii* in humans. In 1890, Eugen Bostroem isolated the causative organism from a culture of grain, grasses, and soil. After Bostroem's discovery there was a general misconception that actinomycosis was a mycosis that affected individuals who chewed grass or straw. The pathogen is still known as the "great masquerader". Bergey classified the organism as bacterial in 1939, but the disease remained classified as a fungus in the 1955 edition of the Control of Communicable Diseases in Man.

Violinist Joseph Joachim died of actinomycosis on 15 August 1907.

Other animals

Actinomycosis occurs rarely in humans but rather frequently in cattle as a disease called *lumpy jaw*. This name refers to the large abscesses that grow on the head and neck of the infected animal. It can also affect swine, horses, and dogs, and less often wild animals and sheep.

Leprosy

Leprosy, also known as Hansen's disease (HD), is a long-term infection caused by the bacilli *Mycobacterium leprae* and *Mycobacterium lepromatosis*. Initially, infections are without symptoms and typically remain this way for 5 to 20 years. Symptoms that develop include granulomas of the nerves, respiratory tract, skin, and eyes. This may result in a lack of ability to feel pain and thus loss of parts of extremities due to repeated injuries or infection due to unnoticed wounds. Weakness and poor eyesight may also be present.

Leprosy is spread between people. This is believed to occur through a cough or contact with fluid from the nose of an infected person. Leprosy occurs more commonly among those living in poverty. Contrary to popular belief, it is not highly contagious. The two main types of disease are based on the number of bacteria present: paucibacillary and multibacillary. The two types are differentiated by the number of poorly pigmented, numb skin patches present, with paucibacillary having five or fewer and multibacillary having more than five. The diagnosis is confirmed by finding acid-fast bacilli in a biopsy of the skin or by detecting the DNA using polymerase chain reaction.

Leprosy is curable with a treatment known as multidrug therapy. Treatment for paucibacillary leprosy is with the medications dapsone and rifampicin for six months. Treatment for multibacillary leprosy consists of rifampicin, dapsone, and clofazimine for 12 months. A number of other antibiotics may also be used. These treatments are provided free of charge by the World Health Organization. Globally in 2012, the number of chronic cases of leprosy was 189,000, down from some 5.2 million in the 1980s. The number of new cases was 230,000. Most new cases occur in 16 countries, with India accounting for more than half. In the past 20 years, 16 million people worldwide have been cured of leprosy. About 200 cases are reported per year in the United States.

Leprosy has affected humanity for thousands of years. The disease takes its name from the Latin word *lepra*, which means "scaly", while the term "Hansen's disease" is named after the physician Gerhard Armauer Hansen. Separating people by placing them in leper colonies still occurs in places such as India, China, and Africa. However, most colonies have closed since leprosy is not very contagious. Social stigma has been associated with leprosy for much of history, which continues to be a barrier to self-reporting and early treatment. Some consider the word "leper" offensive, preferring the phrase "person affected with leprosy". World Leprosy Day was started in 1954 to draw awareness to those affected by leprosy.

Signs and Symptoms

Leprosy is primarily a granulomatous disease of the peripheral nerves and mucosa of the upper respiratory tract; skin lesions (light or dark patches) are the primary external sign. If untreated, leprosy can progress and cause permanent damage to the skin, nerves, limbs, and eyes. Contrary to folklore, leprosy does not cause body parts to fall off, although they can become numb or diseased as a result of secondary infections; these occur as a result of the body's defenses being compromised by the primary disease. Secondary infections, in turn, can result in tissue loss, causing fingers and toes to become shortened and deformed, as cartilage is absorbed into the body.

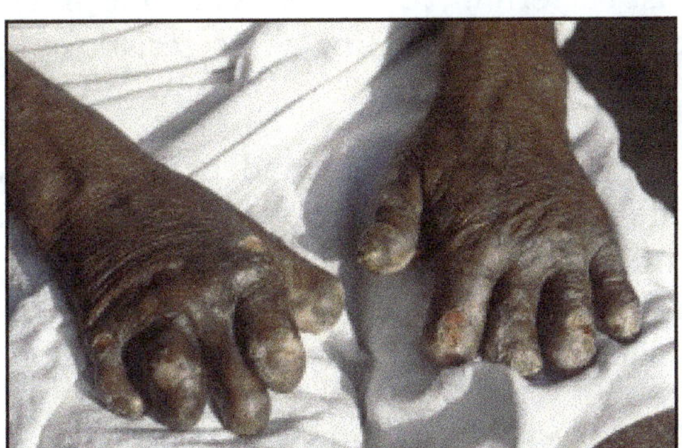

Hands deformed by leprosy.

Leprosy in Tahiti, *circa* 1895.

Cause

M. leprae

M. leprae and *M. lepromatosis* are the causative agents of leprosy. *M. lepromatosis* is a relatively newly identified mycobacterium isolated from a fatal case of diffuse lepromatous leprosy in 2008.

M. leprae, one of the causative agents of leprosy: As an acid-fast bacterium, *M. leprae* appears red when a Ziehl-Neelsen stain is used.

An intracellular, acid-fast bacterium, *M. leprae* is aerobic and rod-shaped, and is surrounded by the waxy cell membrane coating characteristic of the *Mycobacterium* genus.

Due to extensive loss of genes necessary for independent growth, *M. leprae* and *M. lepromatosis* are obligate intracellular pathogens, and unculturable in the laboratory, a factor that leads to difficulty in definitively identifying the organism under a strict interpretation of Koch's postulates. The use of nonculture-based techniques such as molecular genetics has allowed for alternative establishment of causation.

While the causative organisms have to date been impossible to culture *in vitro,* it has been possible to grow them in animals such as mice and armadillos.

Naturally occurring infection also has been reported in nonhuman primates, including the African chimpanzee, sooty mangabey, and cynomolgus macaque, as well as in armadillos and red squirrels.

Risk Factors

The greatest risk factor for developing leprosy is contact with another case of leprosy. Contacts of people with leprosy are five to eight times more likely to develop leprosy than members of the general population. Other risk factors are poorly understood. However, conditions that reduce immune function, such as malnutrition, other illnesses, or host genetic differences, may increase the risk of developing leprosy. Despite this, infection with HIV does not appear to increase the risk of developing leprosy.

Transmission

Transmission of leprosy occurs during close contact with those who are infected. Transmission is believed to be by nasal droplets.

Leprosy is not known to be either sexually transmitted or highly infectious. People are no longer infectious after as little as two weeks of treatment.

Leprosy may also be transmitted to humans by armadillos and may be present in three species of nonhuman primates.

Two exit routes of *M. leprae* from the human body often described are the skin and the nasal mucosa, although their relative importance is not clear. Lepromatous cases show large numbers of organisms deep in the dermis, but whether they reach the skin surface in sufficient numbers is doubtful.

The skin and the upper respiratory tract are most likely entry route. While older research dealt with the skin route, recent research has increasingly favored the respiratory route. Experimental transmission of leprosy through aerosols containing *M. leprae* in immunosuppressed mice was accomplished, suggesting a similar possibility in humans.

Genetics

Name	Locus	OMIM	Gene
LPRS1	10p13	609888	
LPRS2	6q25	607572	*PARK2, PACRG*
LPRS3	4q32	246300	*TLR2*
LPRS4	6p21.3	610988	*LTA*
LPRS5	4p14	613223	*TLR1*
LPRS6	13q14.11	613407	

Several genes have been associated with a susceptibility to leprosy. Many people's immune systems are able to eliminate leprosy during the early infection stage before severe symptoms develop. A defect in cell-mediated immunity may cause susceptibility to leprosy. The region of DNA responsible for this variability is also involved in Parkinson's disease, giving rise to current speculation that the two disorders may be linked in some way at the biochemical level. Some evidence indicates not all people who are infected with *M. leprae* develop leprosy, and genetic factors have long been thought to play a role, due to the observation of clustering of leprosy around certain families, and the failure to understand why certain individuals develop lepromatous leprosy while others develop other types of leprosy.

Pathophysiology

How the infection produces the symptoms of the disease is not known.

Diagnosis

According to the World Health Organization, diagnosis in areas where people are frequently infected is based on one of these main signs:

- Skin lesion consistent with leprosy and with definite sensory loss.
- Positive skin smears.

Skin lesions can be single or multiple, and usually hypopigmented, although occasionally reddish or copper-colored. The lesions may be macules (flat), papules (raised), or nodular. Sensory loss at the skin lesion is important because this feature can help differentiate it from other causes of skin lesions such as tinea versicolor. Thickened nerves are associated with leprosy and can be accompanied by loss of sensation or muscle weakness. However, without the characteristic skin lesion and sensory loss, muscle weakness is not considered a reliable sign of leprosy

Positive skin smears: In some case, acid-fast leprosy bacilli are considered diagnostic; however, the diagnosis is clinical.

Diagnosis in areas where the disease is uncommon, such as the United States, is often delayed because healthcare providers are unaware of leprosy and its symptoms. Early diagnosis and treatment prevent nerve involvement, the hallmark of leprosy, and the disability it causes.

Many kinds of leprosy are known, but some symptoms are common, including runny nose, dry scalp, eye problems, skin lesions, muscle weakness, reddish skin, smooth, shiny, diffuse thickening of facial skin, ear, and hand, loss of sensation in fingers and toes, thickening of peripheral nerves, and flat nose due to destruction of nasal cartilage. Also, phonation and resonation of sound occurs during speech. Often, atrophy of the testes and impotency happen.

Classification

Several different approaches for classifying leprosy exist, but parallels exist.

- The World Health Organization system distinguishes "paucibacillary" and "multibacillary" based upon the proliferation of bacteria.("pauci-" refers to a low quantity).
- The SHAY scale provides five gradations.
- The ICD-10, though developed by the WHO, uses Ridley-Jopling and not the WHO system. It also adds an indeterminate ("I") entry.
- In MeSH, three groupings are used.

WHO	Ridley-Jopling	ICD-10	MeSH	Description	Lepromin test
Paucibacillary	tuberculoid ("TT"), borderline tuberculoid ("BT")	A30.1, A30.2	Tuberculoid	It is characterized by one or more hypopigmented skin macules and patches where skin sensations are lost because of damaged peripheral nerves that have been attacked by the human host's immune cells.	Positive
Multibacillary	midborderline or borderline ("BB")	A30.3	Borderline	Borderline leprosy is of intermediate severity and is the most common form. Skin lesions resemble tuberculoid leprosy, but are more numerous and irregular; large patches may affect a whole limb, and peripheral nerve involvement with weakness and loss of sensation is common. This type is unstable and may become more like lepromatous leprosy or may undergo a reversal reaction, becoming more like the tuberculoid form.	
Multibacillary	borderline lepromatous ("BL"), and lepromatous ("LL")	A30.4, A30.5	Lepromatous	It is associated with symmetric skin lesions, nodules, plaques, thickened dermis, and frequent involvement of the nasal mucosa resulting in nasal congestion and nose bleeds, but, typically, detectable nerve damage is late.	Negative

A difference in immune response to the tuberculoid and lepromatous forms is seen.

Leprosy may also be divided into:

- Early and indeterminate leprosy.

- Tuberculoid leprosy.
- Borderline tuberculoid leprosy.
- Borderline leprosy.
- Borderline lepromatous leprosy.
- Lepromatous leprosy.
- Histoid leprosy.
- Diffuse leprosy of Lucio and Latapí.

This disease may also occur with only neural involvement, without skin lesions.

Prevention

Early detection of the disease is important, since physical and neurological damage may be irreversible even if cured. Medications can decrease the risk of those living with people with leprosy from acquiring the disease and likely those with whom people with leprosy come into contact outside the home. However, concerns are known of resistance, cost, and disclosure of a person's infection status when doing follow-up of contacts. Therefore, the WHO recommends that people who live in the same household be examined for leprosy and only be treated if symptoms are present.

The Bacillus Calmette–Guérin (BCG) vaccine offers a variable amount of protection against leprosy in addition to tuberculosis. It appears to be 26 to 41% effective (based on controlled trials) and about 60% effective based on observational studies with two doses possibly working better than one. Development of a more effective vaccine is ongoing.

Treatment

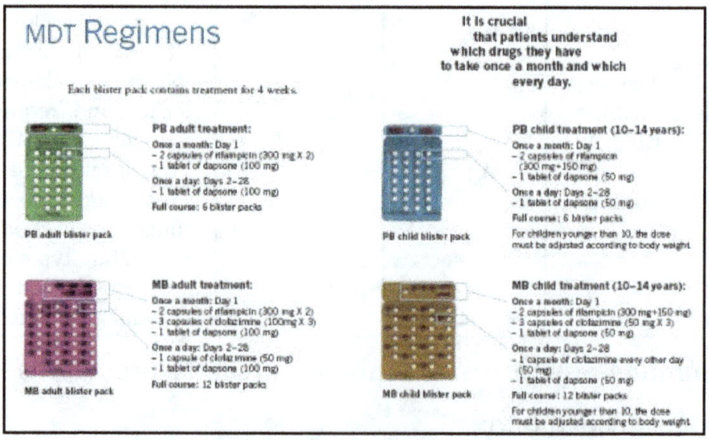

MDT antileprosy drugs: standard regimens.

A number of leprostatic agents are available for treatment. For paucibacillary (PB or tuberculoid) cases, treatment with daily dapsone and monthly rifampicin for six months is recommended. While for multibacillary (MB or lepromatous) cases, treatment with daily dapsone and clofazimine along with monthly rifampicin for 12 months is recommended.

Multidrug therapy (MDT) remains highly effective, and people are no longer infectious after the first monthly dose. It is safe and easy to use under field conditions due to its presentation in calendar blister packs. Relapse rates remain low, and no resistance to the combined drugs is seen.

Epidemiology

In 2012, the number of cases of leprosy was about 180,000. In 2011, the approximate number of new leprosy cases diagnosed was 220,000.

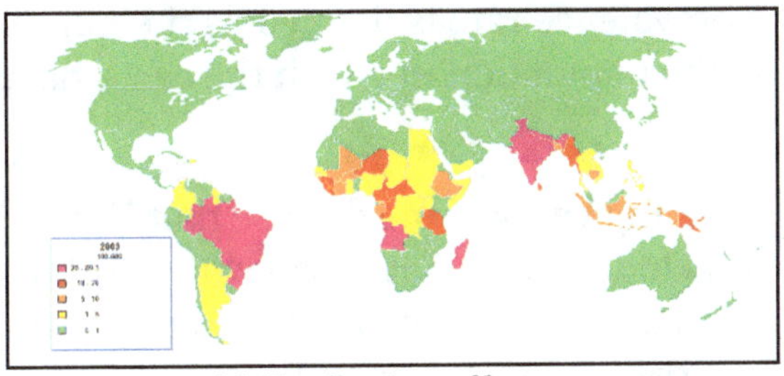

World distribution of leprosy, 2003.

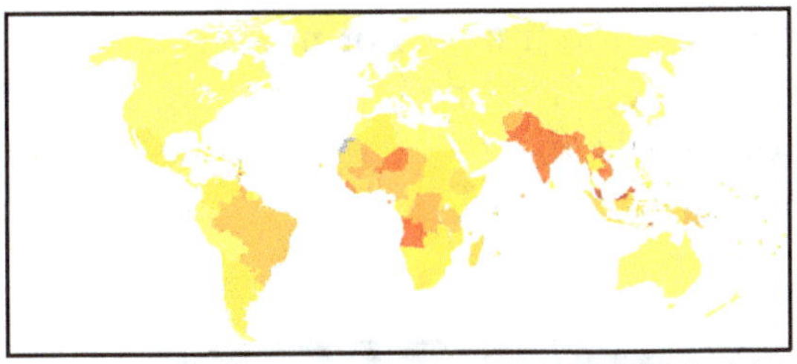

Disability-adjusted life year for leprosy per 100,000 inhabitants in 2004.

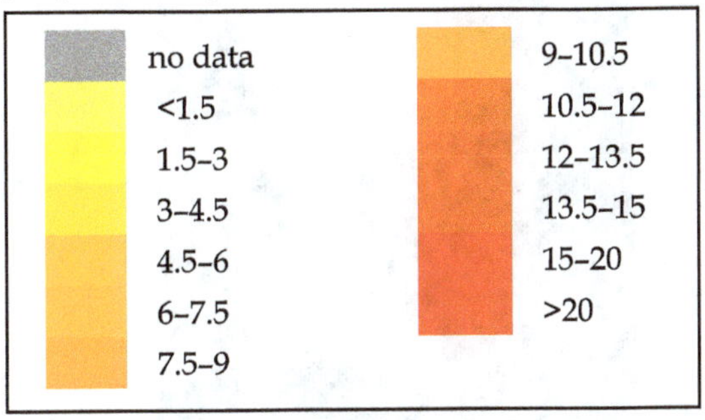

As of 2013, 14 countries contain 95% of the globally reported leprosy cases. Of these, India has the greatest number of cases (59%), followed by Brazil (14%) and Indonesia (8%). Although the number of cases worldwide continues to fall, pockets of high prevalence remain in certain areas such as Brazil, South Asia (India, Nepal, Bhutan), some parts of Africa (Tanzania, Madagascar, Mozambique), and the western Pacific.

The number of cases of leprosy was in the tens of millions in the 1960s, a series of national (the International Federation of Anti-Leprosy Associations) and international (the WHO's "Global Strategy for Reducing Disease Burden Due to Leprosy") initiatives have reduced the total number and the number of new cases of the disease. In 1995, two to three million people were estimated to be permanently disabled because of leprosy.

Disease Burden

Although the number of new leprosy cases occurring each year is important as a measure of transmission, it is difficult to measure due to leprosy's long incubation period, delays in diagnosis after onset of the disease, and the lack of laboratory tools to detect it in the very early stages. Instead, the registered prevalence is used. Registered prevalence is a useful proxy indicator of the disease burden, as it reflects the number of active leprosy cases diagnosed with the disease and receiving treatment with MDT at a given point in time. The prevalence rate is defined as the number of cases registered for MDT treatment among the population in which the cases have occurred, again at a given point in time.

New case detection is another indicator of the disease that is usually reported by countries on an annual basis. It includes cases diagnosed with onset of disease in the year in question (true incidence) and a large proportion of cases with onset in previous years (termed a backlog prevalence of undetected cases).

Endemic countries also report the number of new cases with established disabilities at the time of detection, as an indicator of the backlog prevalence. Determination of the time of onset of the disease is, in general, unreliable, is very labor-intensive, and is seldom done in recording these statistics.

History

G. H. A. Hansen, discoverer of *M. leprae*.

Using comparative genomics, in 2005, geneticists traced the origins and worldwide distribution of leprosy from East Africa or the Near East along human migration routes. They found four strains of *M. leprae* with specific regional locations. Strain 1 occurs predominately in Asia, the Pacific region, and East Africa; strain 4, in West Africa and the Caribbean; strain 3 in Europe, North Africa, and the Americas; and strain 2 only in Ethiopia, Malawi, Nepal/north India, and New Caledonia.

On the basis of this, they offer a map of the dissemination of leprosy in the world. This confirms the spread of the disease along the migration, colonisation, and slave trade routes taken from East Africa to India, West Africa to the New World, and from Africa into Europe and vice versa.

The oldest skeletal evidence for the disease was found in the human remains from the archaeological sites of Balathal and Harappa, in India and Pakistan, respectively.

Although retrospectively identifing descriptions of leprosy-like symptoms is difficult, what appears to be leprosy was discussed by Hippocrates in 460 BC. In 1846, Francis Adams produced *The Seven Books of Paulus Aegineta* which included a commentary on all medical and surgical knowledge and descriptions and remedies to do with leprosy from the Romans, Greeks, and Arabs.

Interpretations of the presence of leprosy have been made on the basis of descriptions in ancient Indian (Atharva Verda and Kausika Sutra), Greek, and Middle Eastern documentary sources that describe skin afflictions.

Skeletal remains from the second millennium BC, discovered in 2009, represent the oldest documented evidence for leprosy. Located at Balathal, in Rajasthan, northwest India, the discoverers suggest that if the disease did migrate from Africa, to India, during the third millennium BC "at a time when there was substantial interaction among the Indus Civilization, Mesopotamia, and Egypt, there needs to be additional skeletal and molecular evidence of leprosy in India and Africa so as to confirm the African origin of the disease." A proven human case was verified by DNA taken from the shrouded remains of a man discovered in a tomb next to the Old City of Jerusalem dated by radiocarbon methods to 1–50 AD.

The causative agent of leprosy, *M. leprae*, was discovered by G. H. Armauer Hansen in Norway in 1873, making it the first bacterium to be identified as causing disease in humans. The first effective treatment (promin) became available in the 1940s. In the 1950s, dapsone was introduced. The search for further effective antileprosy drugs led to the use of clofazimine and rifampicin in the 1960s and 1970s. Later, Indian scientist Shantaram Yawalkar and his colleagues formulated a combined therapy using rifampicin and dapsone, intended to mitigate bacterial resistance. MDT combining all three drugs was first recommended by the WHO] in 1981. These three antileprosy drugs are still used in the standard MDT regimens.

Leprosy was once believed to be highly contagious and was treated with mercury—all of which applied to syphilis, which was first described in 1530. Many early cases thought to be leprosy could actually have been syphilis. Resistance has developed to initial treatment. Until the introduction of MDT in the early 1980s, the disease could not be diagnosed and treated successfully within the community.

Japan still has sanatoriums (although Japan's sanatoriums no longer have active leprosy cases, nor are survivors held in them by law).

The importance of the nasal mucosa in the transmission of *M leprae* was recognized as early as 1898 by Schäffer, in particular that of the ulcerated mucosa.

Society and Culture

Two lepers denied entrance to town, 14th century.

India

India was one of the first countries to have acted against leprosy. India enacted the Leprosy Act of 1898 which institutionalized those affected and segregated them by gender to prevent reproduction. The Act was difficult to enforce, but was only repealed in 1983 after MDT therapy became widely available. In 1983, the National Leprosy Elimination Programme, previously the National Leprosy Control Programme, changed its methods from surveillance to treatment of people with leprosy. India still accounts for over half of the global disease burden.

Treatment Cost

Between 1995 and 1999, the WHO, with the aid of the Nippon Foundation, supplied all endemic countries with free MDT in blister packs, channelled through ministries of health. This free provision was extended in 2000 and again in 2005 with donations by the MDT manufacturer Novartis through the WHO. In the latest agreement signed between the company and the WHO in October 2010, the provision of free MDT by the WHO to all endemic countries has run until the end of 2015. At the national level, nongovernment organizations affiliated with the national programme will continue to be provided with an appropriate free supply of this WHO-supplied MDT by the government.

Historical Texts

Written accounts of leprosy date back thousands of years. Various skin diseases translated as leprosy appear in the ancient Indian text, the *Atharava Veda,* as early as 2000 BC. Another Indian text, the *Laws of Manu* (1500 BC), prohibited contact with those infected with the disease and made marriage to a person infected with leprosy punishable.

Many English translations of the Bible translate *tzaraath* as "leprosy," a confusion that derives from the use of the koine cognate (which can mean any disease causing scaly skin) in the Septuagint. Ancient sources such as the Talmud (Sifra 63) make clear that *tzaraath* refers to various types of lesions or stains associated with ritual impurity and occurring on cloth, leather, or houses, as well as skin. It may sometimes be a symptom of the disease described in this article, but has many other causes, as well. The New Testament describes instances of Jesus healing people with leprosy (Luke 5:10), although the precise relationship between this, *tzaraath,* and Hansen's disease is not established.

The biblical perception that people with leprosy were unclean may be connected to a passage from Leviticus 13: 44-46. Judeo-Christian belief held that leprosy was a moral disease, and early Christians believed that those affected by leprosy were being punished by God for sinful behavior. Moral associations have persisted throughout history. Pope Gregory the Great (540-604) and Isidor of Seville (560-636) considered people with the disease to be heretics.

Middle Ages

It is believed that a rise in leprosy in Europe occurred in the Middle Ages based on the increased number of hospitals created to treat leprosy patients in the 12th and 13th centuries. France alone had nearly 2,000 leprosariums during this period.

Medieval leper bell.

The social perception in medieval communities was generally one of fear, and those people infected with the disease were thought to be unclean, untrustworthy, and morally corrupt. People with leprosy were also often required to wear clothing that identified them as such or carry a bell announcing their presence. Segregation from mainstream society was common. The third Lateran Council of 1179 and a 1346 edict by King

Edward expelled lepers from city limits. Because of the moral stigma of the disease, methods of treatment were both physical and spiritual, and leprosariums were established under the purview of the church.

Nineteenth Century

Norway

Norway was the location of a progressive stance on leprosy tracking and treatment, and played an influential role in European understanding of the disease. In 1832, Dr. JJ Hjort conducted the first leprosy survey, thus establishing a basis for epidemiological surveys. Subsequent surveys resulted in the establishment of a national leprosy registry to study the causes of leprosy and for tracking of rate of infection.

Early leprosy research throughout Europe was conducted by Norwegian scientists, Daniel Cornelius Danielssen and C.W. Boeck. Their work resulted in the establishment of the National Leprosy Research and Treatment Center. Danielssen and Boeck believed the cause of leprosy transmission was hereditary. This stance was influential in advocating for the isolation of those infected by gender to prevent reproduction.

Colonialism and Imperialism

Though leprosy in Europe was again on the decline by the 1860s, Western countries embraced isolation treatment out of fear of the spread of disease from developing countries, minimal understanding of bacteriology, lack of diagnostic ability or knowledge of how contagious the disease was, and missionary activity. Growing imperialism and pressures of the industrial revolution resulted in a Western presence in countries where leprosy was endemic, namely the British presence in India. Isolation treatment methods were observed by Surgeon-Mayor Henry Vandyke Carter of the British Colony in India while visiting Norway, and these methods were applied in India with the financial and logistical assistance of religious missionaries. Colonial and religious influence and associated stigma continued to be a major factor in the treatment and public perception of leprosy in endemic developing countries until the mid-20th century.

Stigma

Despite effective treatment and education efforts, leprosy stigma continues to be problematic in endemic developing countries. Leprosy is most prevalent amongst impoverished or marginalized populations where social stigma is likely to be compounded by other social inequities. Fears of ostracism, loss of employment, or expulsion from family and society may contribute to a delayed diagnosis and treatment.

Folk models of belief, lack of education, and religious connotations of the disease continue to influence social perceptions of those afflicted in many parts of the world. In Brazil, for example, folklore holds that leprosy is transmitted by dogs, it is a disease associated with sexual promiscuity, and is sometimes thought to be punishment for sins or moral transgressions. Socioeconomic factors also have a direct impact. Lower-class domestic workers who are often employed by those in a higher socioeconomic class may find their employment in jeopardy as physical manifestations of the disease become apparent. Skin discoloration and darker pigmentation resulting from the disease also has social repercussions.

In extreme cases in northern India, leprosy is equated with an "untouchable" status that "often persists long after (individuals with leprosy) have been cured of the disease, creating lifelong prospects of divorce, eviction, loss of employment, and ostracism from family and social networks."

Programs and Treatment

The WHO states that diagnosis and treatment with MDT is easy and effective, and a 45% decline in disease burden has occurred since MDT has become more widely available. The organization emphasizes the importance of fully integrating leprosy treatment into public health services, effective diagnosis and treatment, and access to information.

In some instances in India, community based rehabilitation is embraced by local governments and NGOs alike. Often, the identity cultivated by a community environment is preferable to reintegration, and models of self-management and collective agency independent of NGOs and government support have been desirable and successful.

Acute Prostatitis

Acute prostatitis is a serious bacterial infection of the prostate gland. This infection is a medical emergency. It should be distinguished from other forms of prostatitis such as chronic bacterial prostatitis and chronic pelvic pain syndrome (CPPS).

Signs and Symptoms

Men with this disease often have chills, fever, pain in the lower back and genital area, urinary frequency and urgency often at night, burning or painful urination, body aches, and a demonstrable infection of the urinary tract, as evidenced by white blood cells and bacteria in the urine. Acute prostatitis may be a complication of prostate biopsy. Often, the prostate gland is very tender to palpation through the rectum.

Diagnosis

Acute prostatitis is relatively easy to diagnose due to its symptoms that suggest infection. The organism may be found in blood or urine, and sometimes in both. Common bacteria are *Escherichia coli, Klebsiella, Proteus, Pseudomonas, Enterobacter,*

Enterococcus, Serratia, and *Staphylococcus aureus*. This can be a medical emergency in some patients and hospitalization with intravenous antibiotics may be required. A complete blood count reveals increased white blood cells. Sepsis from prostatitis is very rare, but may occur in immunocompromised patients; high fever and malaise generally prompt blood cultures, which are often positive in sepsis. A prostate massage should never be done in a patient with suspected acute prostatitis, since it may induce sepsis. Since bacteria causing the prostatitis is easily recoverable from the urine, prostate massage is not required to make the diagnosis. Rectal palpation usually reveals an enlarged, exquisitely tender, swollen prostate gland, which is firm, warm, and, occasionally, irregular to the touch. C-reactive protein is elevated in most cases.

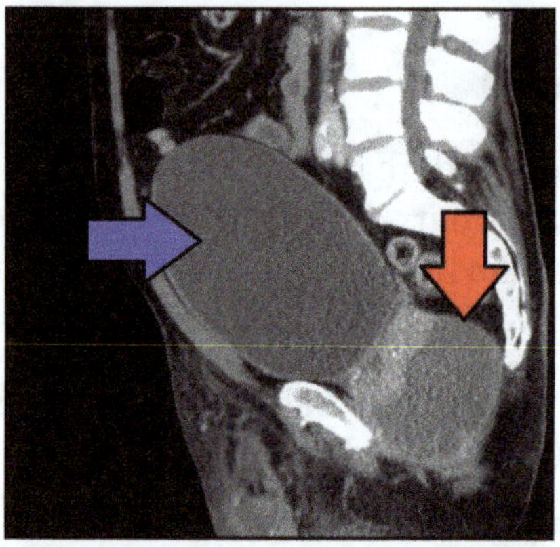

Abscess of the prostate resulting in urinary retention.

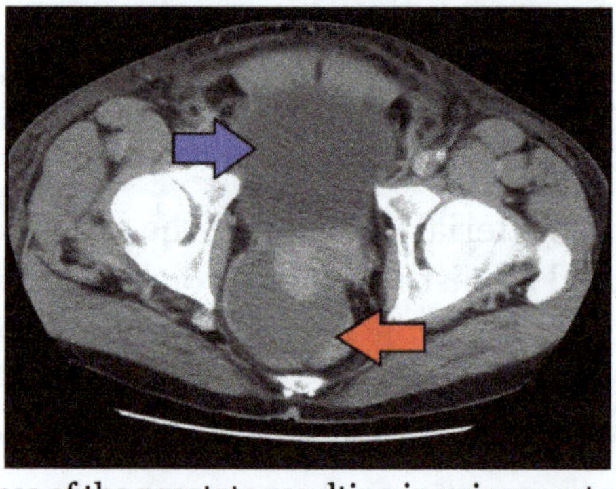

Abscess of the prostate resulting in urinary retention.

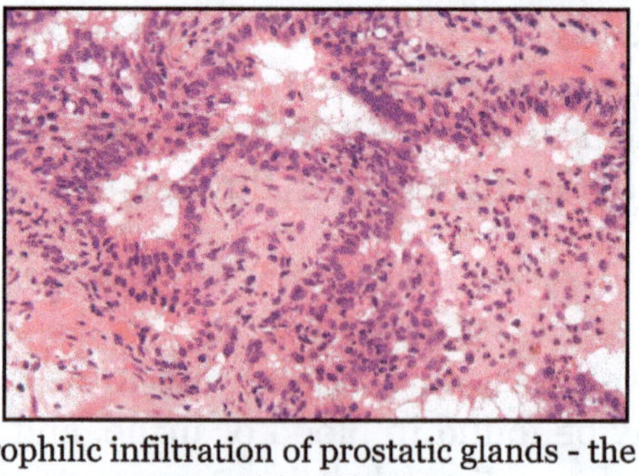

Micrograph showing a neutrophilic infiltration of prostatic glands - the histologic correlate of acute prostatitis. H&E stain.

Prostate biopsies are not indicated as the (clinical) features (described above) are diagnostic. The histologic correlate of acute prostatitis is a neutrophilic infiltration of the prostate gland.

Acute prostatitis is associated with a transiently elevated PSA, i.e., the PSA is increased during an episode of acute prostatitis and then decreases again after it has resolved. PSA testing is not indicated in the context of uncomplicated acute prostatitis.

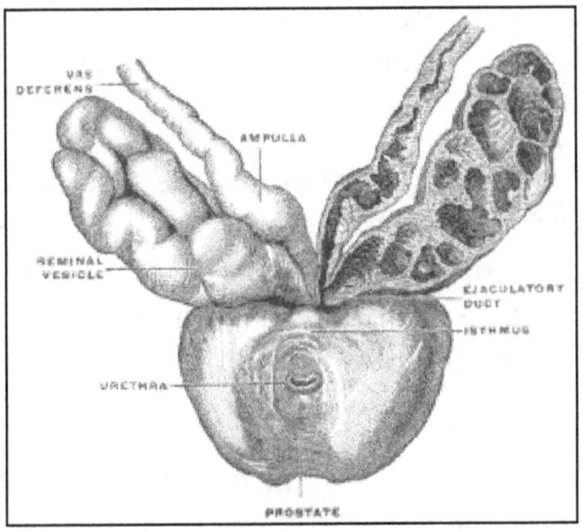

Prostate, urethra, and seminal vesicles.

Treatment

Antibiotics are the first line of treatment in acute prostatitis (Cat. I). Antibiotics usually resolve acute prostatitis infections in a very short time, however a minimum of two to four weeks of therapy is recommended to eradicate the offending organism completely. Appropriate antibiotics should be used, based on the microbe causing the infection. Some antibiotics have very poor penetration of the prostatic capsule, others, such as Ciprofloxacin, Co-trimoxazole and tetracyclines such as doxycycline penetrate well. In acute prostatitis, penetration of the prostate is not as important as for category II because the intense inflammation disrupts the prostate-blood barrier. It is more important to choose a bacteriocidal antibiotic (kills bacteria, e.g. quinolone) rather than a bacteriostatic antibiotic (slows bacterial growth, e.g. tetracycline) for acute potentially life-threatening infections. Severely ill patients may need hospitalization, while nontoxic patients can be treated at home with bed rest, analgesics, stool softeners, and hydration. Patients in urinary retention are best managed with a suprapubic catheter or intermittent catheterization. Lack of clinical response to antibiotics should raise the suspicion of an abscess and prompt an imaging study such as a transrectal ultrasound (TRUS).

Prognosis

Full recovery without sequelae is usual.

Tuberculosis

Tuberculosis (TB) is an infectious disease caused by the bacterium *Mycobacterium tuberculosis* (MTB). Tuberculosis generally affects the lungs, but can also affect other parts of the body. Most infections do not have symptoms, known as latent tuberculosis. About 10% of latent infections progress to active disease which, if left untreated, kills about half of those infected. The classic symptoms of active TB are a chronic cough with blood-containing sputum, fever, night sweats, and weight loss. The historical term "consumption" came about due to the weight loss. Infection of other organs can cause a wide range of symptoms.

Tuberculosis is spread through the air when people who have active TB in their lungs cough, spit, speak, or sneeze. People with latent TB do not spread the disease. Active infection occurs more often in people with HIV/AIDS and in those who smoke. Diagnosis of active TB is based on chest X-rays, as well as microscopic examination and culture of body fluids. Diagnosis of latent TB relies on the tuberculin skin test (TST) or blood tests.

Prevention of TB involves screening those at high risk, early detection and treatment of cases, and vaccination with the bacillus Calmette-Guérin vaccine. Those at high risk include household, workplace, and social contacts of people with active TB. Treatment requires the use of multiple antibiotics over a long period of time. Antibiotic resistance is a growing problem with increasing rates of multiple drug-resistant tuberculosis (MDR-TB).

One-third of the world's population is thought to be infected with TB. New infections occur in about 1% of the population each year. In 2014, there were 9.6 million cases of active TB which resulted in 1.5 million deaths. More than 95% of deaths occurred in developing countries. The number of new cases each year has decreased since 2000. About 80% of people in many Asian and African countries test positive while 5–10% of people in the United States population tests positive by the tuberculin test. Tuberculosis has been present in humans since ancient times.

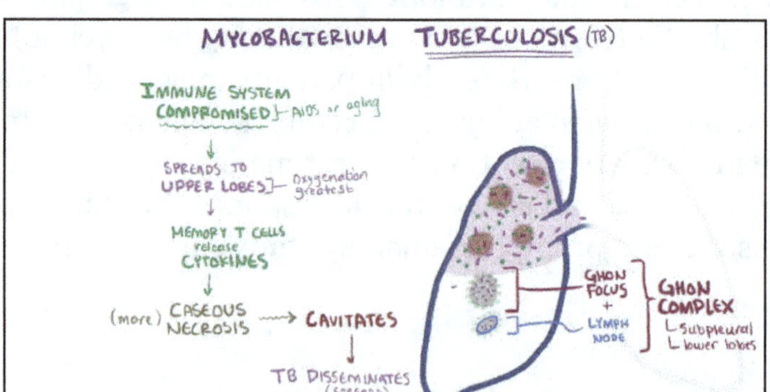

Video explanation.

Signs and Symptoms

Tuberculosis may infect any part of the body, but most commonly occurs in the lungs (known as pulmonary tuberculosis). Extrapulmonary TB occurs when tuberculosis develops outside of the lungs, although extrapulmonary TB may coexist with pulmonary TB.

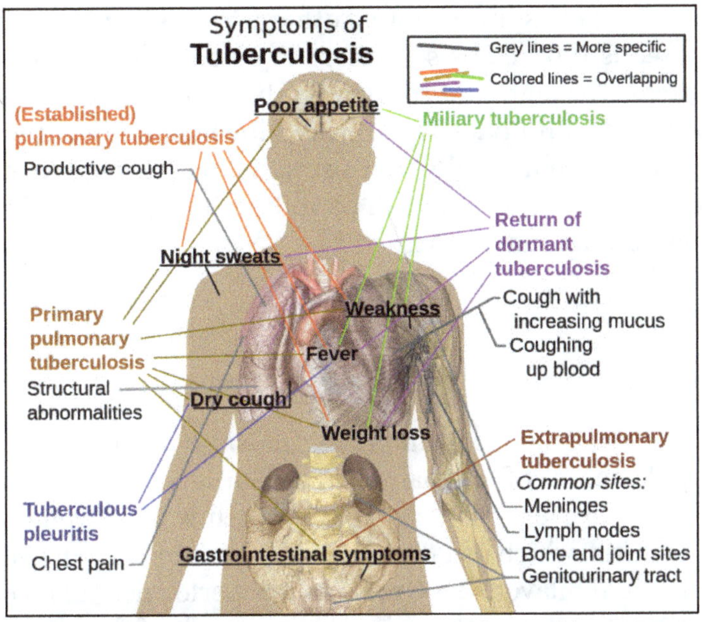

The main symptoms of variants and stages of tuberculosis are given, with many symptoms overlapping with other variants, while others are more (but not entirely) specific for certain variants. Multiple variants may be present simultaneously.

General signs and symptoms include fever, chills, night sweats, loss of appetite, weight loss, and fatigue. Significant nail clubbing may also occur.

Pulmonary

If a tuberculosis infection does become active, it most commonly involves the lungs (in about 90% of cases). Symptoms may include chest pain and a prolonged cough producing sputum. About 25% of people may not have any symptoms (i.e. they remain "asymptomatic"). Occasionally, people may cough up blood in small amounts, and in very rare cases, the infection may erode into the pulmonary artery or a Rasmussen's aneurysm, resulting in massive bleeding. Tuberculosis may become a chronic illness and cause extensive scarring in the upper lobes of the lungs. The upper lung lobes are more frequently affected by tuberculosis than the lower ones. The reason for this difference is not clear. It may be due to either better air flow, or poor lymph drainage within the upper lungs.

Extrapulmonary

In 15–20% of active cases, the infection spreads outside the lungs, causing other kinds of TB. These are collectively denoted as "extrapulmonary tuberculosis".

Extrapulmonary TB occurs more commonly in immunosuppressed persons and young children. In those with HIV, this occurs in more than 50% of cases. Notable extrapulmonary infection sites include the pleura (in tuberculous pleurisy), the central nervous system (in tuberculous meningitis), the lymphatic system (in scrofula of the neck), the genitourinary system (in urogenital tuberculosis), and the bones and joints (in Pott disease of the spine), among others. When it spreads to the bones, it is also known as "osseous tuberculosis", a form of osteomyelitis. Sometimes, bursting of a tubercular abscess through skin results in tuberculous ulcer. An ulcer originating from nearby infected lymph nodes is painless, slowly enlarging and has an appearance of "wash leather". A potentially more serious, widespread form of TB is called "disseminated tuberculosis", also known as miliary tuberculosis. Miliary TB makes up about 10% of extrapulmonary cases.

Causes

Mycobacteria

The main cause of TB is *Mycobacterium tuberculosis*, a small, aerobic, nonmotile bacillus. The high lipid content of this pathogen accounts for many of its unique clinical characteristics. It divides every 16 to 20 hours, which is an extremely slow rate compared with other bacteria, which usually divide in less than an hour. Mycobacteria have an outer membrane lipid bilayer. If a Gram stain is performed, MTB either stains very weakly "Gram-positive" or does not retain dye as a result of the high lipid and mycolic acid content of its cell wall. MTB can withstand weak disinfectants and survive in a dry state for weeks. In nature, the bacterium can grow only within the cells of a host organism, but *M. tuberculosis* can be cultured in the laboratory.

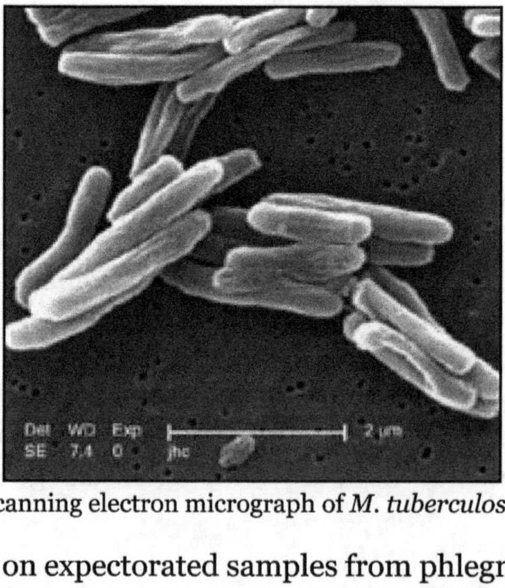

Scanning electron micrograph of *M. tuberculosis*.

Using histological stains on expectorated samples from phlegm (also called "sputum"), scientists can identify MTB under a microscope. Since MTB retains certain stains even after being treated with acidic solution, it is classified as an acid-fast bacillus. The most common acid-fast staining techniques are the Ziehl–Neelsen stain and the Kinyoun stain, which dye acid-fast bacilli a bright red that stands out against a blue background. Auramine-rhodamine staining and fluorescence microscopy are also used.

The *M. tuberculosis* complex (MTBC) includes four other TB-causing mycobacteria: *M. bovis*, *M. africanum*, *M. canetti*, and *M. microti*. *M. africanum* is not widespread, but it is a significant cause of tuberculosis in parts of Africa. *M. bovis* was once a common cause of tuberculosis, but the introduction of pasteurized milk has almost completely eliminated this as a public health problem in developed countries. *M. canetti* is rare and seems to be limited to the Horn of Africa, although a few cases have been seen in African emigrants. *M. microti* is also rare and is seen almost only in immunodeficient people, although its prevalence may be significantly underestimated.

Other known pathogenic mycobacteria include *M. leprae*, *M. avium*, and *M. kansasii*. The latter two species are classified as "nontuberculous mycobacteria" (NTM). NTM cause neither TB nor leprosy, but they do cause pulmonary diseases that resemble TB.

Risk Factors

A number of factors make people more susceptible to TB infections. The most important risk factor globally is HIV; 13% of all people with TB are infected by the virus. This is a particular problem in sub-Saharan Africa, where rates of HIV are high. Of people without HIV who are infected with tuberculosis, about 5–10% develop active disease during their lifetimes; in contrast, 30% of those coinfected with HIV develop the active disease.

Tuberculosis is closely linked to both overcrowding and malnutrition, making it one of the principal diseases of poverty. Those at high risk thus include: people who inject illicit drugs, inhabitants and employees of locales where vulnerable people gather (e.g. prisons and homeless shelters), medically underprivileged and resource-poor communities, high-risk ethnic minorities, children in close contact with high-risk category patients, and health-care providers serving these patients.

Chronic lung disease is another significant risk factor. Silicosis increases the risk about 30-fold. Those who smoke cigarettes have nearly twice the risk of TB compared to nonsmokers.

Other disease states can also increase the risk of developing tuberculosis. These include alcoholism and diabetes mellitus (three-fold increase).

Certain medications, such as corticosteroids and infliximab (an anti-αTNF monoclonal antibody), are becoming increasingly important risk factors, especially in the developed world.

Genetic susceptibility also exists, for which the overall importance remains undefined.

Mechanism

Public health campaigns in the 1920s tried to halt the spread of TB.

Transmission

When people with active pulmonary TB cough, sneeze, speak, sing, or spit, they expel infectious aerosol droplets 0.5 to 5.0 μm in diameter. A single sneeze can release up to 40,000 droplets. Each one of these droplets may transmit the disease, since the infectious dose of tuberculosis is very small (the inhalation of fewer than 10 bacteria may cause an infection).

People with prolonged, frequent, or close contact with people with TB are at particularly high risk of becoming infected, with an estimated 22% infection rate. A person with active but untreated tuberculosis may infect 10–15 (or more) other people per year. Transmission should occur from only people with active TB – those with latent infection are not thought to be contagious. The probability of transmission from one person to another depends upon several factors, including the number of infectious droplets expelled by the carrier, the effectiveness of ventilation, the duration of exposure, the virulence of the *M. tuberculosis* strain, the level of immunity in the uninfected person, and others. The cascade of person-to-person spread can be circumvented by segregating those with active ("overt") TB and putting them on anti-TB drug regimens. After about two weeks of effective treatment, subjects with nonresistant active infections generally do not remain contagious to others. If someone does become infected, it typically takes three to four weeks before the newly infected person becomes infectious enough to transmit the disease to others.

Pathogenesis

About 90% of those infected with *M. tuberculosis* have asymptomatic, latent TB infections (sometimes called LTBI), with only a 10% lifetime chance that the latent infection will progress to overt, active tuberculous disease. In those with HIV, the risk of developing active TB increases to nearly 10% a year. If effective treatment is not given, the death rate for active TB cases is up to 66%.

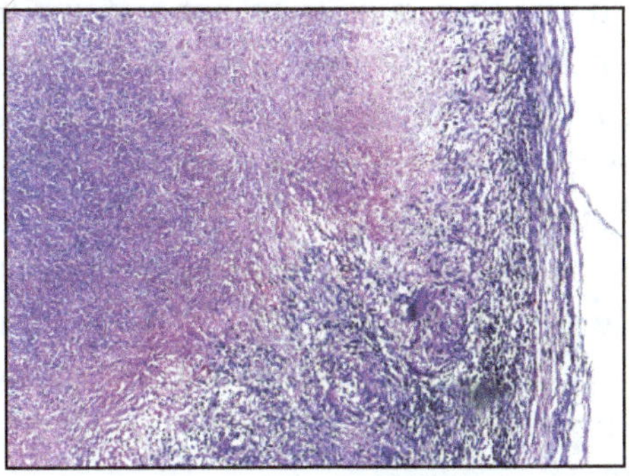

Microscopy of tuberculous epididymitis. H&E stain.

TB infection begins when the mycobacteria reach the pulmonary alveoli, where they invade and replicate within endosomes of alveolar macrophages. Macrophages identify the bacterium as foreign and attempt to eliminate it by phagocytosis. During this process, the bacterium is enveloped by the macrophage and stored temporarily in a membrane-bound vesicle called a phagosome. The phagosome then combines with a lysosome to create a phagolysosome. In the phagolysosome, the cell attempts to use reactive oxygen species and acid to kill the bacterium. However, *M. tuberculosis* has a thick, waxy mycolic acid capsule that protects it from these toxic substances. *M. tuberculosis* is able to reproduce inside the macrophage and will eventually kill the immune cell.

The primary site of infection in the lungs, known as the "Ghon focus", is generally located in either the upper part of the lower lobe, or the lower part of the upper lobe. Tuberculosis of the lungs may also occur via infection from the blood stream. This is known as a Simon focus and is typically found in the top of the lung. This hematogenous transmission can also spread infection to more distant sites, such as peripheral lymph nodes, the kidneys, the brain, and the bones. All parts of the body can be affected by the disease, though for unknown reasons it rarely affects the heart, skeletal muscles, pancreas, or thyroid.

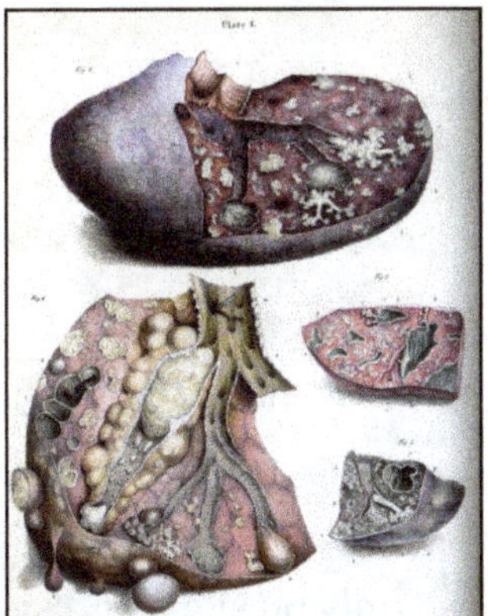

Robert Carswell's illustration of tubercle.

Tuberculosis is classified as one of the granulomatous inflammatory diseases. Macrophages, T lymphocytes, B lymphocytes, and fibroblasts aggregate to form granulomas, with lymphocytes surrounding the infected macrophages. When other macrophages attack the infected macrophage, they fuse together to form a giant multinucleated cell in the alveolar lumen. The granuloma may prevent dissemination of the mycobacteria and provide a local environment for interaction of cells of the immune system. However, more recent evidence suggests that the bacteria use the granulomas to avoid destruction by the host's immune system. Macrophages and dendritic cells in the granulomas are unable to present antigen to lymphocytes; thus the immune response is suppressed. Bacteria inside the granuloma can become dormant, resulting in latent infection. Another feature of the granulomas is the development of abnormal cell death (necrosis) in the center of tubercles. To the naked eye, this has the texture of soft, white cheese and is termed caseous necrosis.

If TB bacteria gain entry to the blood stream from an area of damaged tissue, they can spread throughout the body and set up many foci of infection, all appearing as tiny, white tubercles in the tissues. This severe form of TB disease, most common in young children and those with HIV, is called miliary tuberculosis. People with this disseminated TB have a high fatality rate even with treatment (about 30%).

In many people, the infection waxes and wanes. Tissue destruction and necrosis are often balanced by healing and fibrosis. Affected tissue is replaced by scarring and cavities filled with caseous necrotic material. During active disease, some of these cavities are joined to the air passages bronchi and this material can be coughed up. It contains living bacteria, so can spread the infection. Treatment with appropriate antibiotics kills bacteria and allows healing to take place. Upon cure, affected areas are eventually replaced by scar tissue.

Diagnosis

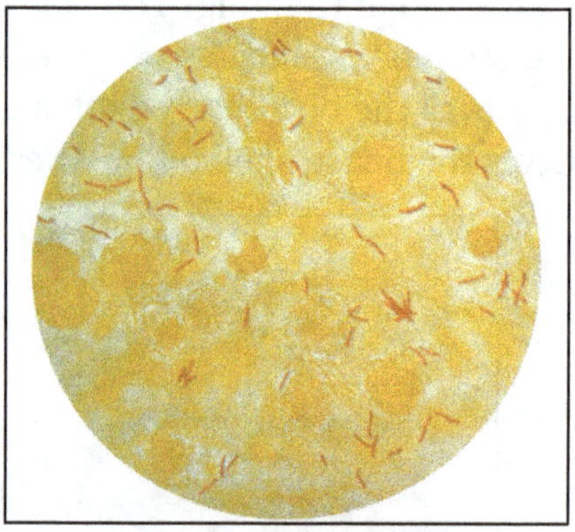

M. tuberculosis (stained red) in sputum.

Active Tuberculosis

Diagnosing active tuberculosis based only on signs and symptoms is difficult, as is diagnosing the disease in those who are immunosuppressed. A diagnosis of TB should, however, be considered in those with signs of lung disease or constitutional symptoms lasting longer than two weeks. A chest X-ray and multiple sputum cultures for acid-fast bacilli are typically part of the initial evaluation. Interferon-γ release assays and tuberculin skin tests are of little use in the developing world. IGRA have similar limitations in those with HIV.

A definitive diagnosis of TB is made by identifying *M. tuberculosis* in a clinical sample (e.g., sputum, pus, or a tissue biopsy). However, the difficult culture process for this slow-growing organism can take two to six weeks for blood or sputum culture. Thus, treatment is often begun before cultures are confirmed.

Nucleic acid amplification tests and adenosine deaminase testing may allow rapid diagnosis of TB. These tests, however, are not routinely recommended, as they rarely alter how a person is treated. Blood tests to detect antibodies are not specific or sensitive, so they are not recommended.

Latent Tuberculosis

The Mantoux tuberculin skin test is often used to screen people at high risk for TB. Those who have been previously immunized may have a false-positive test result. The test may be falsely negative in those with sarcoidosis, Hodgkin's lymphoma, malnutrition, and most notably, active tuberculosis. Interferon gamma release assays (IGRAs), on a blood sample, are recommended in those who are positive to the Mantoux test. These are not affected by immunization or most environmental mycobacteria, so they generate fewer false-positive results. However, they are affected by *M. szulgai*, *M. marinum*, and *M. kansasii*. IGRAs may increase sensitivity when used in addition to the skin test, but may be less sensitive than the skin test when used alone.

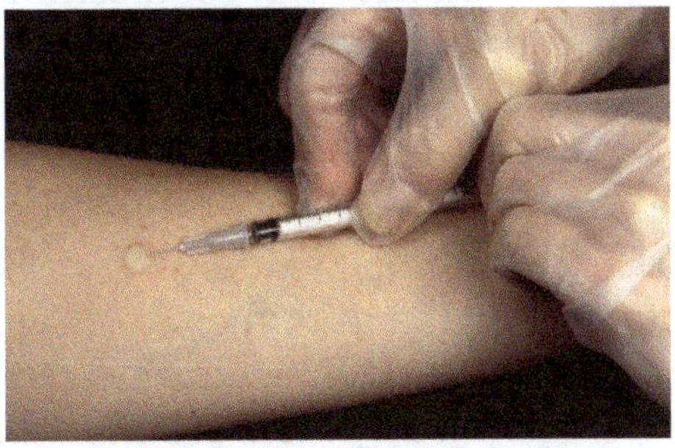

Mantoux tuberculin skin test.

Prevention

Tuberculosis prevention and control efforts rely primarily on the vaccination of infants and the detection and appropriate treatment of active cases. The World Health Organization has achieved some success with improved treatment regimens, and a small decrease in case numbers. The US Preventive Services Task Force (USPSTF) recommends screening people who are at high risk for latent tuberculosis with either tuberculin skin tests or interferon-gamma release assays.

Vaccines

The only available vaccine as of 2011 is Bacillus Calmette-Guérin (BCG). In children it decreases the risk of getting the infection by 20% and the risk of infection turning into disease by nearly 60%.

It is the most widely used vaccine worldwide, with more than 90% of all children being vaccinated. The immunity it induces decreases after about ten years. As tuberculosis is uncommon in most of Canada, the United Kingdom, and the United States, BCG is administered to only those people at high risk. Part of the reasoning against the use of the vaccine is that it makes the tuberculin skin test falsely positive, reducing the test's use in screening. A number of new vaccines are currently in development.

Public Health

The World Health Organization declared TB a "global health emergency" in 1993, and in 2006, the Stop TB Partnership developed a Global Plan to Stop Tuberculosis that aimed to save 14 million lives between its launch and 2015. A number of targets they set were not achieved by 2015, mostly due to the increase in HIV-associated tuberculosis and the emergence of multiple drug-resistant tuberculosis. A tuberculosis classification system developed by the American Thoracic Society is used primarily in public health programs.

Management

Treatment of TB uses antibiotics to kill the bacteria. Effective TB treatment is difficult, due to the unusual structure and chemical composition of the mycobacterial cell wall, which hinders the entry of drugs and makes many antibiotics ineffective. The two antibiotics most commonly used are isoniazid and rifampicin, and treatments can be prolonged, taking several months. Latent TB treatment usually employs a single antibiotic, while active TB disease is best treated with combinations of several antibiotics to reduce the risk of the bacteria developing antibiotic resistance. People with latent infections are also treated to prevent them from progressing to active TB disease later in life. Directly observed therapy, i.e., having a health care provider watch the person take their medications, is recommended by the WHO in an effort to reduce the number of people not appropriately taking antibiotics. The evidence to support this practice over people simply taking their medications independently is poor. Methods to remind people of the importance of treatment do, however, appear effective.

New Onset

The recommended treatment of new-onset pulmonary tuberculosis, as of 2010, is six months of a combination of antibiotics containing rifampicin, isoniazid, pyrazinamide, and ethambutol for the first two months, and only rifampicin and isoniazid for the last four months. Where resistance to isoniazid is high, ethambutol may be added for the last four months as an alternative.

Recurrent Disease

If tuberculosis recurs, testing to determine to which antibiotics it is sensitive is important before determining treatment. If multiple drug-resistant TB is detected, treatment with at least four effective antibiotics for 18 to 24 months is recommended.

Medication Resistance

Primary resistance occurs when a person becomes infected with a resistant strain of TB. A person with fully susceptible MTB may develop secondary (acquired) resistance during therapy because of inadequate treatment, not taking the prescribed regimen appropriately (lack of compliance), or using low-quality medication. Drug-resistant TB is a serious public health issue in many developing countries, as its treatment is longer and requires more expensive drugs. MDR-TB is defined as resistance to the two most effective first-line TB drugs: rifampicin and isoniazid. Extensively drug-resistant TB is also resistant to three or more of the six classes of second-line drugs. Totally drug-resistant TB is resistant to all currently used drugs. It was first observed in 2003 in Italy, but not widely reported until 2012, and has also been found in Iran and India. Bedaquiline is tentatively supported for use in multiple drug-resistant TB.

XDR-TB is a term sometimes used to define *extensively resistant* TB, and constitutes one in ten cases of MDR-TB. Cases of XDR TB have been identified in more than 90% of countries.

Prognosis

Progression from TB infection to overt TB disease occurs when the bacilli overcome the immune system defenses and begin to multiply. In primary TB disease (some 1–5% of cases), this occurs soon after the initial infection. However, in the majority of cases, a latent infection occurs with no obvious symptoms. These dormant bacilli produce active tuberculosis in 5–10% of these latent cases, often many years after infection.

The risk of reactivation increases with immunosuppression, such as that caused by infection with HIV. In people coinfected with *M. tuberculosis* and HIV, the risk of reactivation increases to 10% per year. Studies using DNA fingerprinting of *M. tuberculosis* strains have shown reinfection contributes more substantially to recurrent TB than previously thought, with estimates that it might account for more than 50% of reactivated cases in areas where TB is common. The chance of death from a case of tuberculosis is about 4% as of 2008, down from 8% in 1995.

Epidemiology

Roughly one-third of the world's population has been infected with *M. tuberculosis*, with new infections occurring in about 1% of the population each year. However, most infections with *M. tuberculosis* do not cause TB disease, and 90–95% of infections remain asymptomatic. In 2012, an estimated 8.6 million chronic cases were active. In 2010, 8.8 million new cases of TB were diagnosed, and 1.20–1.45 million deaths occurred, most of these occurring in developing countries. Of these 1.45 million deaths, about 0.35 million occur in those also infected with HIV.

Tuberculosis is the second-most common cause of death from infectious disease (after those due to HIV/AIDS). The total number of tuberculosis cases has been decreasing since 2005, while new cases have decreased since 2002. China has achieved particularly dramatic progress, with about an 80% reduction in its TB mortality rate between 1990 and 2010. The number of new cases has declined by 17% between 2004–2014. Tuberculosis is more common in developing countries; about 80% of the population in many Asian and African countries test positive in tuberculin tests, while only 5–10% of the US population test positive. Hopes of totally controlling the disease have been dramatically dampened because of a number of factors, including the difficulty of developing an effective vaccine, the expensive and time-consuming diagnostic process, the necessity of many months of treatment, the increase in HIV-associated tuberculosis, and the emergence of drug-resistant cases in the 1980s.

In 2007, the country with the highest estimated incidence rate of TB was Swaziland, with 1,200 cases per 100,000 people. India had the largest total incidence, with an estimated 2.0 million new cases. In developed countries, tuberculosis is less common and is found mainly in urban areas. Rates per 100,000 people in different areas of the world were: globally 178, Africa 332, the Americas 36, Eastern Mediterranean 173, Europe 63, Southeast Asia 278, and Western Pacific 139 in 2010. In Canada and Australia, tuberculosis is many times more common among the aboriginal peoples, especially in remote areas. In the United States Native Americans have a fivefold greater mortality from TB, and racial and ethnic minorities accounted for 84% of all reported TB cases.

The rates of TB varies with age. In Africa, it primarily affects adolescents and young adults. However, in countries where incidence rates have declined dramatically (such as the United States), TB is mainly a disease of older people and the immunocompromised (risk factors are listed above). Worldwide, 22 "high-burden" states or countries together experience 80% of cases as well as 83% of deaths.

History

Tuberculosis has been present in humans since antiquity. The earliest unambiguous detection of *M. tuberculosis* involves evidence of the disease in the remains of bison in Wyoming dated to around 17,000 years ago. However, whether tuberculosis originated in bovines, then was transferred to humans, or whether it diverged from a common ancestor, is currently unclear. A comparison of the genes of *M. tuberculosis* complex (MTBC) in humans to MTBC in animals suggests humans did not acquire MTBC from animals during animal domestication, as was previously believed. Both strains of the tuberculosis bacteria share a common ancestor, which could have infected humans even before the Neolithic Revolution.

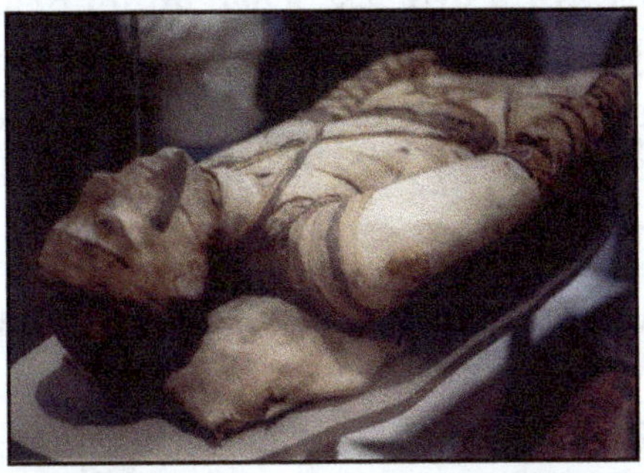

Egyptian mummy in the British Museum – tubercular decay has been found in the spine.

Skeletal remains show prehistoric humans (4000 BC) had TB, and researchers have found tubercular decay in the spines of Egyptian mummies dating from 3000–2400 BC. Genetic studies suggest TB was present in the Americas from about 100 AD.

Before the Industrial Revolution, folklore often associated tuberculosis with vampires. When one member of a family died from it, the other infected members would lose their health slowly. People believed this was caused by the original person with TB draining the life from the other family members.

Although the pulmonary form associated with tubercles was established as a pathology by Dr Richard Morton in 1689, due to the variety of its symptoms, TB was not identified as a single disease until the 1820s. It was not named "tuberculosis" until 1839, by J. L. Schönlein.

During 1838–1845, Dr. John Croghan, the owner of Mammoth Cave, brought a number of people with tuberculosis into the cave in the hope of curing the disease with the constant temperature and purity of the cave air; they died within a year. Hermann Brehmer opened the first TB sanatorium in 1859 in Görbersdorf (now Sokołowsko), Silesia.

Dr. Robert Koch discovered the tuberculosis bacillus.

The bacillus causing tuberculosis, *M. tuberculosis*, was identified and described on 24 March 1882 by Robert Koch. He received the Nobel Prize in physiology or medicine in 1905 for this discovery. Koch did not believe the bovine (cattle) and human tuberculosis diseases were similar, which delayed the recognition of infected milk as a source of infection. Later, the risk of transmission from this source was dramatically reduced by the invention of the pasteurization process. Koch announced a glycerine extract of the tubercle bacilli as a "remedy" for tuberculosis in 1890, calling it "tuberculin". While it was not effective, it was later successfully adapted as a screening test for the presence of pre-symptomatic tuberculosis. The World Tuberculosis Day was established on 24 March for this reason.

Albert Calmette and Camille Guérin achieved the first genuine success in immunization against tuberculosis in 1906, using attenuated bovine-strain tuberculosis. It was called bacille Calmette–Guérin (BCG). The BCG vaccine was first used on humans in 1921 in France, but received widespread acceptance in the US, Great Britain, and Germany only after World War II.

Tuberculosis caused the most widespread public concern in the 19th and early 20th centuries as an endemic disease of the urban poor. In 1815, one in four deaths in England was due to "consumption". By 1918, one in six deaths in France was still caused by TB. After TB was determined to be contagious, in the 1880s, it was put on a notifiable disease list in Britain; campaigns were started to stop people from spitting in public places, and the infected poor were "encouraged" to enter sanatoria that resembled prisons (the sanatoria for the middle and upper classes offered excellent care and constant medical attention). Whatever the (purported) benefits of the "fresh air" and labor in the sanatoria, even under the best conditions, 50% of those who entered died within five years (*circa* 1916).

In Europe, rates of tuberculosis began to rise in the early 1600s to a peak level in the 1800s, when it caused nearly 25% of all deaths. By the 1950s, mortality had decreased nearly 90%. Improvements in public health began significantly reducing rates of tuberculosis even before the arrival of streptomycin and other antibiotics, although the disease remained a significant threat to public health such that when the Medical Research Council was formed in Britain in 1913, its initial focus was tuberculosis research.

In 1946, the development of the antibiotic streptomycin made effective treatment and cure of TB a reality. Prior to the introduction of this drug, the only treatment (except sanatoria) was surgical intervention, including the "pneumothorax technique", which involved collapsing an infected lung to "rest" it and allow tuberculous lesions to heal.

Because of the emergence of MDR-TB, surgery has been re-introduced as an option within the generally accepted standard of care in treating TB infections. Current surgical interventions involve removal of pathological chest cavities ("bullae") in the lungs to reduce the number of bacteria and to increase the exposure of the remaining bacteria to drugs in the bloodstream, thereby simultaneously reducing the total bacterial load and increasing the effectiveness of systemic antibiotic therapy.

Hopes of completely eliminating TB (*cf.* smallpox) from the population were dashed after the rise of drug-resistant strains in the 1980s. The subsequent resurgence of tuberculosis resulted in the declaration of a global health emergency by the World Health Organization in 1993.

Society and Culture

Names

Phthisis (Φθισις) is a Greek word for consumption, an old term for pulmonary tuberculosis; around 460 BC, Hippocrates described phthisis as a disease of dry seasons. The abbreviation "TB" is short for *tubercle bacillus*.

"Consumption" was the most common nineteenth century English word for the disease. The Latin root "con" meaning "completely" is linked to "sumere" meaning "to take up from under." In The Life and Death of Mr. Badman by John Bunyan, the author calls consumption "the captain of all these men of death."

Public Health Efforts

The World Health Organization, Bill and Melinda Gates Foundation, and US government are subsidizing a fast-acting diagnostic tuberculosis test for use in low- and middle-income countries. In addition to being fast-acting, the test can determine if there is resistance to the antibiotic rifampicin which may indicate multi-drug resistant tuberculosis and is accurate in those who are also infected with HIV. Many resource-poor places as of 2011 have access to only sputum microscopy.

India had the highest total number of TB cases worldwide in 2010, in part due to poor disease management within the private and public health care sector. Programs such as the Revised National Tuberculosis Control Program are working to reduce TB levels amongst people receiving public health care.

A 2014 the EIU-healthcare report that the need to address apathy and urging for increased funding. The report cites among others Lucica Ditui "[TB] is like an orphan. It has been neglected even in countries with a high burden and often forgotten by donors and those investing in health interventions."

Slow progress has led to frustration, expressed by the executive director of the Global Fund to Fight AIDS, Tuberculosis and Malaria – Mark Dybul: "we have the tools to end TB as a pandemic and public health threat on the planet, but we are not doing it." Several international organizations are pushing for more transparency in treatment, and more countries are implementing mandatory reporting of cases to the government, although adherence is often sketchy. Commercial treatment providers may at times overprescribe second-line drugs as well as supplementary treatment, promoting demands for further regulations. The government of Brazil provides universal TB-care, which reduces this problem. Conversely, falling rates of TB-infection may not relate to the number of programs directed at reducing infection rates, but may be tied to increased level of education, income, and health of the population. Costs of the disease, as calculated by the World Bank in 2009 may exceed 150 billion USD per year in "high burden" countries. Lack of progress eradicating the disease may also be due to lack of patient follow-up – as among the 250M rural migrants in China.

Stigma

Slow progress in preventing the disease may in part be due to stigma associated with TB. Stigma may be due to the fear of transmission from affected individuals. This stigma may additionally arise due to links between TB and poverty, and in Africa, AIDS. Such stigmatization may be both real and perceived, for example; in Ghana individuals with TB are banned from attending public gatherings.

Stigma towards TB may result in delays in seeking treatment, lower treatment compliance, and family members keeping cause of death secret – allowing the disease to spread further. At odds is Russia, where stigma was associated with increased treatment compliance. TB stigma also affects socially marginalized individuals to a greater degree and varies between regions.

One way to decrease stigma may be through the promotion of "TB clubs", where those infected may share experiences and offer support, or through counseling. Some studies have shown TB education programs to be effective in decreasing stigma, and may thus be effective in increasing treatment adherence. Despite this, studies on the relationship between reduced stigma and mortality are lacking as of 2010, and similar efforts to decrease stigma surrounding AIDS have been minimally effective. Some have claimed the stigma to be worse than the disease, and healthcare providers may unintentionally reinforce stigma, as those with TB are often perceived as difficult or otherwise undesirable. A greater understanding of the social and cultural dimensions of tuberculosis may also help with stigma reduction.

Research

The BCG vaccine has limitations, and research to develop new TB vaccines is ongoing. A number of potential candidates are currently in phase I and II clinical trials. Two main approaches are being used to attempt to improve the efficacy of available vaccines. One approach involves adding a subunit vaccine to BCG, while the other strategy is attempting to create new and better live vaccines. MVA85A, an example of a subunit vaccine, currently in trials in South Africa, is based on a genetically modified vaccinia virus. Vaccines are hoped to play a significant role in treatment of both latent and active disease.

To encourage further discovery, researchers and policymakers are promoting new economic models of vaccine development, including prizes, tax incentives, and advance market commitments. A number of groups, including the Stop TB Partnership, the South African Tuberculosis Vaccine Initiative, and the Aeras Global TB Vaccine Foundation, are involved with research. Among these, the Aeras Global TB Vaccine Foundation received a gift of more than $280 million (US) from the Bill and Melinda Gates Foundation to develop and license an improved vaccine against tuberculosis for use in high burden countries.

A number of medications are being studied for multi drug resistant tuberculosis including: bedaquiline and delamanid. Bedaquiline received U.S. Food and Drug Administration (FDA) approval in late 2012. The safety and effectiveness of these new agents are still uncertain, because they are based on the results of a relatively small studies. However, existing data suggest that patients taking bedaquiline in addition to standard TB therapy are five times more likely to die than those without the new drug, which has resulted in medical journal articles raising health policy questions about why the FDA approved the drug and whether financial ties to the company making bedaquiline influenced physicians' support for its use.

Other Animals

Mycobacteria infect many different animals, including birds, rodents, and reptiles. The subspecies *Mycobacterium tuberculosis*, though, is rarely present in wild animals. An effort to eradicate bovine tuberculosis caused by *Mycobacterium bovis* from the cattle and deer herds of New Zealand has been relatively successful. Efforts in Great Britain have been less successful.

As of 2015, tuberculosis appears to be widespread among captive elephants in the US. It is believed that the animals originally acquired the disease from humans, a process called reverse zoonosis. Because the disease can spread through the air to infect both humans and other animals, it is a public health concern affecting circuses and zoos.

Typhoid Fever

Typhoid fever, also known simply as typhoid, is a bacterial infection due to *Salmonella typhi* that causes symptoms which may vary from mild to severe and usually begin six to thirty days after exposure. Often there is a gradual onset of a high fever over several days. Weakness, abdominal pain, constipation, and headaches also commonly occur. Diarrhea is uncommon and vomiting is not usually severe. Some people develop a skin rash with rose colored spots. In severe cases there may be confusion. Without treatment symptoms may last weeks or months. Other people may carry the bacterium without being affected; however, they are still able to spread the disease to others. Typhoid fever is a type of enteric fever along with paratyphoid fever.

The cause is the bacterium *Salmonella typhi*, also known as *Salmonella enterica* serotype typhi, growing in the intestines and blood. Typhoid is spread by eating or drinking food or water contaminated with the feces of an infected person. Risk factors include poor sanitation and poor hygiene. Those who travel to the developing world are also at risk and only humans can be infected. Diagnosis is by either culturing the bacteria or detecting the bacterium's DNA in the blood, stool, or bone marrow. Culturing the bacterium can be difficult. Bone marrow testing is the most accurate. Symptoms are similar to that of many other infectious diseases. Typhus is a different disease.

A typhoid vaccine can prevent about 30% to 70% of cases during the first two years. The vaccine may have some effect for up to seven years. It is recommended for those at high risk or people traveling to areas where the disease is common. Other efforts to prevent the disease include providing clean drinking water, better sanitation, and better handwashing. Until it has been confirmed that an individual's infection is cleared, the individual should not prepare food for others. Treatment of disease is with antibiotics such as azithromycin, fluoroquinolones or third generation cephalosporins. Resistance to these antibiotics has been developing, which has made treatment of the disease more difficult.

In 2013 there were 11 million new cases reported worldwide. The disease is most common in India. Children are most commonly affected. Rates of disease decreased in the developed world in the 1940s as a result of improved sanitation and use of antibiotics to treat the disease. Each year in the United States about 400 cases are reported and it is estimated that the disease occurs in about 6,000 people. In 2013 it resulted in about 161,000 deaths worldwide - down from 181,000 in 1990 (about 0.3% of the global total). The risk of death may be as high as 20% without treatment. With treatment it is between 1 and 4%. The name typhoid means "resembling typhus" due to the similarity in symptoms.

Signs and Symptoms

Classically, the course of untreated typhoid fever is divided into four distinct stages, each lasting about a week. Over the course of these stages, the patient becomes exhausted and emaciated.

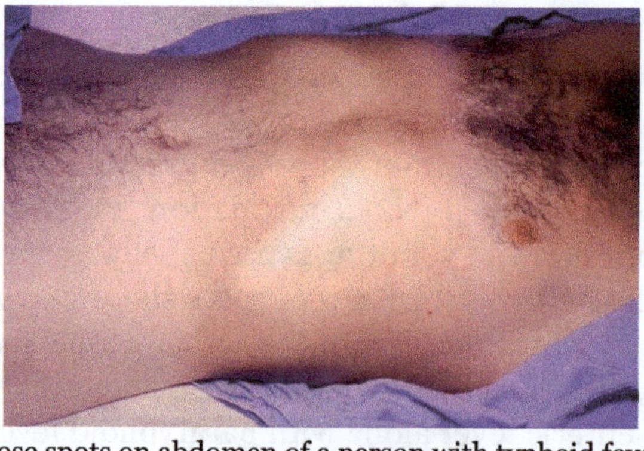

Rose spots on abdomen of a person with typhoid fever.

- In the first week, the body temperature rises slowly, and fever fluctuations are seen with relative bradycardia (Faget sign), malaise, headache, and cough. A bloody nose (epistaxis) is seen in a quarter of cases, and abdominal pain is also possible. A decrease in the number of circulating white blood cells (leukopenia) occurs with eosinopenia and relative lymphocytosis; blood cultures are positive for *Salmonella typhi* or *S. paratyphi*. The Widal test is negative in the first week.
- In the second week, the person is often too tired to get up, with high fever in plateau around 40 °C (104 °F) and bradycardia (sphygmothermic dissociation or Faget sign), classically with a dicrotic pulse wave. Delirium is frequent, often calm, but sometimes agitated. This delirium gives to typhoid the nickname of "nervous fever". Rose spots appear on the lower chest and abdomen in around a third of patients. Rhonchi are heard in lung bases.
- The abdomen is distended and painful in the right lower quadrant, where borborygmi can be heard. Diarrhea can occur in this stage: six to eight stools in a day, green, comparable to pea soup, with a characteristic smell. However, constipation is also frequent. The spleen and liver are enlarged (hepatosplenomegaly) and tender, and liver transaminases are elevated. The Widal test is strongly positive, with antiO and antiH antibodies. Blood cultures are sometimes still positive at this stage.
- (The major symptom of this fever is that the fever usually rises in the afternoon up to the first and second week.)
- In the third week of typhoid fever, a number of complications can occur:
    - Intestinal haemorrhage due to bleeding in congested Peyer's patches; this can be very serious, but is usually not fatal.
    - Intestinal perforation in the distal ileum: this is a very serious complication and is frequently fatal. It may occur without alarming symptoms until septicaemia or diffuse peritonitis sets in.
    - Encephalitis.
    - Respiratory diseases such as pneumonia and acute bronchitis.
    - Neuropsychiatric symptoms (described as "muttering delirium" or "coma vigil"), with picking at bedclothes or imaginary objects.
    - Metastatic abscesses, cholecystitis, endocarditis, and osteitis.
    - The fever is still very high and oscillates very little over 24 hours. Dehydration ensues, and the patient is delirious (typhoid state). One-third of affected individuals develop a macular rash on the trunk.
    - Platelet count goes down slowly and risk of bleeding rises.
- By the end of third week, the fever starts subsiding.

Cause

Transmission

The bacterium that causes typhoid fever may be spread through poor hygiene habits and public sanitation conditions, and sometimes also by flying insects feeding on feces. Public education campaigns encouraging people to wash their hands after defecating and before handling food are an important component in controlling spread of the disease. According to statistics from the United States Centers for Disease Control and Prevention (CDC), the chlorination of drinking water has led to dramatic decreases in the transmission of typhoid fever in the United States.

A 1939 conceptual illustration showing various ways that typhoid bacteria can contaminate a water well (center).

Bacteria

The cause is the bacterium *Salmonella typhi*, also known as *Salmonella enterica* serotype typhi.

There are two main types of Typhi namely the ST1 and ST2 based on MLST subtyping scheme, which are currently widespread globally.

Diagnosis

Diagnosis is made by any blood, bone marrow or stool cultures and with the Widal test (demonstration of antibodies against *Salmonella* antigens O-somatic and H-flagellar). In epidemics and less wealthy countries, after excluding malaria, dysentery, or pneumonia, a therapeutic trial time with chloramphenicol is generally undertaken while awaiting the results of the Widal test and cultures of the blood and stool.

The Widal test is time-consuming, and often, when a diagnosis is reached, it is too late to start an antibiotic regimen.

The term 'enteric fever' is a collective term that refers to severe typhoid and paratyphoid.

Prevention

Sanitation and hygiene are important to prevent typhoid. Typhoid does not affect animals other than humans. Typhoid can only spread in environments where human feces or urine are able to come into contact with food or drinking water. Careful food preparation and washing of hands are crucial to prevent typhoid. Industrialization, and in particular, the invention of the automobile, contributed greatly to the elimination of typhoid fever, as it eliminated the public health hazards associated with having horse manure in the public street which led to large number of flies.

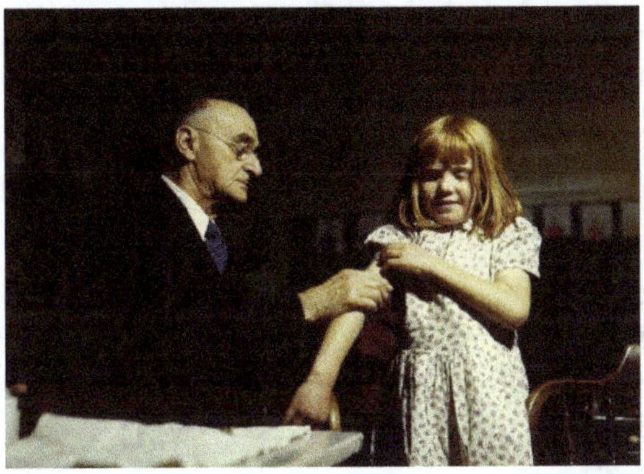

Doctor administering a typhoid vaccination at a school in San Augustine County, Texas, 1943.

Two typhoid vaccines are licensed for use for the prevention of typhoid: the live, oral Ty21a vaccine (sold as Vivotif by Crucell Switzerland AG) and the injectable typhoid polysaccharide vaccine (sold as Typhim Vi by Sanofi Pasteur and 'Typherix by GlaxoSmithKline). Both are recommended for travellers to areas where typhoid is endemic. Boosters are recommended every five years for the oral vaccine and every two years for the injectable form. An older, killed-whole-cell vaccine is still used in countries where the newer preparations are not available, but this vaccine is no longer recommended for use because it has a higher rate of side effects (mainly pain and inflammation at the site of the injection).

Treatment

The rediscovery of oral rehydration therapy in the 1960s provided a simple way to prevent many of the deaths of diarrheal diseases in general.

Where resistance is uncommon, the treatment of choice is a fluoroquinolone such as ciprofloxacin. Otherwise, a third-generation cephalosporin such as ceftriaxone or cefotaxime is the first choice. Cefixime is a suitable oral alternative.

Typhoid fever, when properly treated, is not fatal in most cases. Antibiotics, such as ampicillin, chloramphenicol, trimethoprim-sulfamethoxazole, amoxicillin, and ciprofloxacin, have been commonly used to treat typhoid fever in microbiology. Treatment of the disease with antibiotics reduces the case-fatality rate to about 1%.

When untreated, typhoid fever persists for three weeks to a month. Death occurs in 10% to 30% of untreated cases. In some communities, however, case-fatality rates may reach as high as 47%.

Surgery

Surgery is usually indicated in cases of intestinal perforation. Most surgeons prefer simple closure of the perforation with drainage of the peritoneum. Small-bowel resection is indicated for patients with multiple perforations.

If antibiotic treatment fails to eradicate the hepatobiliary carriage, the gallbladder should be resected. Cholecystectomy is not always successful in eradicating the carrier state because of persisting hepatic infection.

Resistance

As resistance to ampicillin, chloramphenicol, trimethoprim-sulfamethoxazole, and streptomycin is now common, these agents have not been used as first–line treatment of typhoid fever for almost 20 years. Typhoid resistant to these agents is known as multidrug-resistant typhoid (MDR typhoid).

Ciprofloxacin resistance is an increasing problem, especially in the Indian subcontinent and Southeast Asia. Many centres are shifting from using ciprofloxacin as the first line for treating suspected typhoid originating in South America, India, Pakistan, Bangladesh, Thailand, or Vietnam. For these people, the recommended first-line treatment is ceftriaxone. Also, azithromycin has been suggested to be better at treating typhoid in resistant populations than both fluoroquinolone drugs and ceftriaxone. Azithromycin significantly reduces relapse rates compared with ceftriaxone.

A separate problem exists with laboratory testing for reduced susceptibility to ciprofloxacin: current recommendations are that isolates should be tested simultaneously against ciprofloxacin (CIP) and against nalidixic acid (NAL), and that isolates that are sensitive to both CIP and NAL should be reported as "sensitive to ciprofloxacin", but that isolates testing sensitive to CIP but not to NAL should be reported as "reduced sensitivity to ciprofloxacin". However, an analysis of 271 isolates showed that around 18% of isolates with a reduced susceptibility to ciprofloxacin (MIC 0.125–1.0 mg/l) would not be picked up by this method. How this problem can be solved is not certain, because most laboratories around the world (including the West) are dependent on disk testing and cannot test for MICs.

Epidemiology

In 2000, typhoid fever caused an estimated 21.7 million illnesses and 217,000 deaths. It occurs most often in children and young adults between 5 and 19 years old. In 2013 it resulted in about 161,000 deaths – down from 181,000 in 1990. Infants, children, and adolescents in south-central and Southeast Asia experience the greatest burden of illness. Outbreaks of typhoid fever are also frequently reported from sub-Saharan Africa and countries in South-east Asia. Historically, in the pre-antibiotic era, the case fatality rate of typhoid fever was 10–20%. Today, with prompt treatment, it is less than 1%. However, about 3-5% of individuals who are infected will develop a chronic infection in the gall bladder. Since *S.* Typhi is human-restricted, these chronic carriers become the crucial reservoir, which can persist for decades for further spread of the disease, further complicating the identification and treatment of the disease. Lately, the study of Typhi associated with a large outbreak and a carrier at the genome level provides new insights into the pathogenesis of the pathogen.

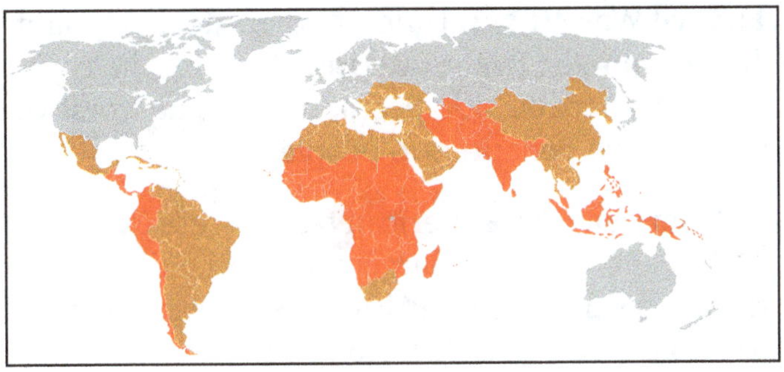

Incidence of typhoid fever
♦ Strongly endemic
♦ Endemic
♦ Sporadic cases

History

In 430 BC, a plague, which some believe to have been typhoid fever, killed one-third of the population of Athens, including their leader Pericles. Following this disaster, the balance of power shifted from Athens to Sparta, ending the Golden Age of Pericles that had marked Athenian dominance in the Greek ancient world. The ancient historian Thucydides also contracted the disease, but he survived to write about the plague. His writings are the primary source on this outbreak, and modern academics and medical scientists consider epidemic typhus the most likely cause. In 2006, a study detected DNA sequences similar to those of the bacterium responsible for typhoid fever in dental pulp extracted from a burial pit dated to the time of the outbreak.

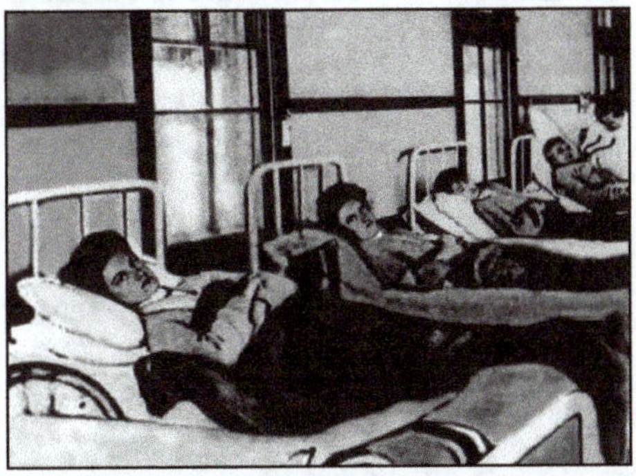

Mary Mallon ("Typhoid Mary") in a hospital bed (foreground): She was forcibly quarantined as a carrier of typhoid fever in 1907 for three years and then again from 1915 until her death in 1938.

The cause of the plague has long been disputed and other scientists have disputed the findings, citing serious methodologic flaws in the dental pulp-derived DNA study. The disease is most commonly transmitted through poor hygiene habits and public sanitation conditions; during the period in question related to Athens above, the whole population of Attica was besieged within the Long Walls and lived in tents.

Some historians believe that English colony of Jamestown, Virginia, died out from typhoid. Typhoid fever killed more than 6000 settlers in the New World between 1607 and 1624.

During the American Civil War, 81,360 Union soldiers died of typhoid or dysentery, far more than died of battle wounds. In the late 19th century, the typhoid fever mortality rate in Chicago averaged 65 per 100,000 people a year. The worst year was 1891, when the typhoid death rate was 174 per 100,000 people.

During the Spanish–American War, American troops were exposed to typhoid fever in stateside training camps and overseas, largely due to inadequate sanitation systems. The Surgeon General of the Army, George Miller Sternberg, suggested that the War Department create a Typhoid Fever Board. Major Walter Reed, Edward O. Shakespeare, and Victor C. Vaughan were appointed August 18, 1898, with Reed being designated the President of the Board. The Typhoid Board determined that during the war, more soldiers died from this disease than from yellow fever or from battle wounds. The Board promoted sanitary measures including latrine policy, disinfection, camp relocation, and water sterilization, but by far the most successful antityphoid method was vaccination, which became compulsory in June 1911 for all federal troops.

The most notorious carrier of typhoid fever, but by no means the most destructive, was Mary Mallon, also known as Typhoid Mary. In 1907, she became the first carrier in the United States to be identified and traced. She was a cook in New York who is closely associated with 53 cases and three deaths. Public health authorities told Mary to give up working as a cook or have her gall bladder removed, as she had a chronic infection that kept her active as a carrier of the disease. Mary quit her job, but returned later under a false name. She was detained and quarantined after another typhoid outbreak. She died of pneumonia after 26 years in quarantine.

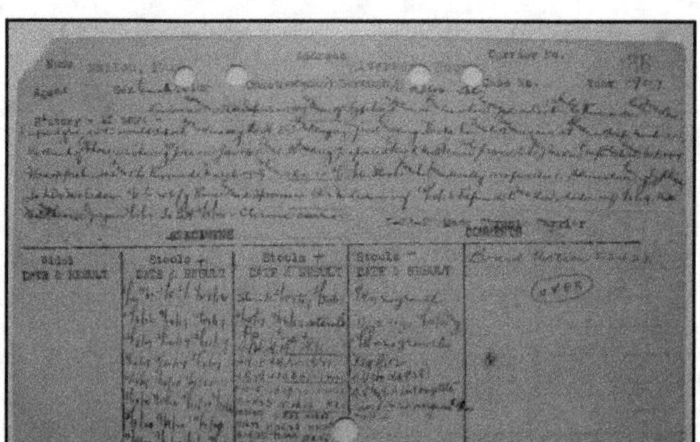

Original stool report for Mary Mallon, 1907.

Development of Vaccination

During the course of treatment of a typhoid outbreak in a local village in 1838, English country doctor William Budd realised the "poisons" involved in infectious diseases multiplied in the intestines of the sick, were present in their excretions, and could be transmitted to the healthy through their consumption of contaminated water. He proposed strict isolation or quarantine as a method for containing such outbreaks in the future. The medical and scientific communities did not identify the role of microorganisms in infectious disease until the work of Louis Pasteur.

Almroth Edward Wright developed the first effective typhoid vaccine.

In 1880, Karl Joseph Eberth described a bacillus that he suspected was the cause of typhoid. In 1884, pathologist Georg Theodor August Gaffky (1850–1918) confirmed Eberth's findings, and the organism was given names such as Eberth's bacillus, *Eberthella typhi*, and Gaffky-Eberth bacillus. Today, the bacillus that causes typhoid fever goes by the scientific name *Salmonella enterica enterica*, serovar Typhi.

The British bacteriologist Almroth Edward Wright first developed an effective typhoid vaccine at the Army Medical School in Netley, Hampshire. It was introduced in 1896 and used successfully by the British during the Boer War in South Africa. At that time, typhoid often killed more soldiers at war than were lost due to enemy combat. Wright further developed his vaccine at a newly opened research department at St Mary's Hospital Medical School in London from 1902, where he established a method for measuring protective substances (opsonin) in human blood.

Citing the example of the Second Boer War, during which many soldiers died from easily preventable diseases, Wright convinced the British Army that 10 million vaccines should be produced for the troops being sent to the Western Front, thereby saving up to half a million lives during World War I. The British Army was the only combatant at the outbreak of the war to have its troops fully immunized against the bacterium. For the first time, their casualties due to combat exceeded those from disease.

In 1909, Frederick F. Russell, a U.S. Army physician, adopted Wright's typhoid vaccine for use with the US Army, and two years later, his vaccination program became the first in which an entire army was immunized. It eliminated typhoid as a significant cause of morbidity and mortality in the U.S. military.

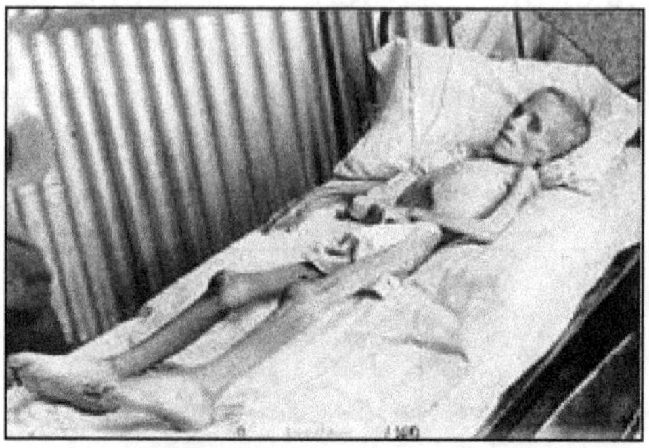

Lizzie van Zyl was a child inmate in a British-run concentration camp in South Africa who died from typhoid fever during the Boer War (1899–1902).

Most developed countries saw declining rates of typhoid fever throughout the first half of the 20th century due to vaccinations and advances in public sanitation and hygiene. In 1908, the chlorination of public drinking water was a significant step in the US in the control of typhoid fever. The first permanent disinfection of drinking water in the U.S. was made to the Jersey City, New Jersey, water supply. Credit for the decision to build the chlorination system has been given to John L. Leal. The chlorination facility was designed by George W. Fuller. In 1942 doctors introduced antibiotics in clinical practice, greatly reducing mortality. Today, the incidence of typhoid fever in developed countries is around five cases per million people per year.

A notable outbreak occurred in Aberdeen, Scotland, in 1964. This was due to contaminated tinned meat sold at the city's branch of the William Low chain of stores. No fatalities resulted.

In 2004–05 an outbreak in the Democratic Republic of Congo resulted in more than 42,000 cases and 214 deaths.

Names

The disease has been referred to by various names, often associated with symptoms, such as gastric fever, enteric fever, abdominal typhus, infantile remittant fever, slow fever, nervous fever, and pythogenic fever.

Bacterial Cold Water Disease

Bacterial cold water disease (BCWD) is a bacterial disease of salmonid fish. It is caused by *Flavobacterium psychrophilum* (previously classified in the genus *Cytophaga*), a gram-negative rod-shaped bacterium of the family Flavobacteriaceae. The disease typically occurs at temperatures below 13°C, and it can be seen in any area with water temperatures consistently below 15°C. Salmon are the most commonly affected species. This disease is not zoonotic.

Asymptomatic carrier fish and contaminated water provide reservoirs for disease. Transmission is mainly horizontal, but vertical transmission can also occur.

BCWD may be referred to by a number of other names including cold water disease, peduncle disease, fit rot, tail rot and rainbow trout fry mortality syndrome.

Diagnosis

Fish infected with typical BCWD have lesions on the skin and fins. Fins may appear dark, torn, split, ragged, frayed and may even be lost completely. Affected fish are often lethargic and stop feeding. Infection may spread systemically. Salmonid fish can also get a chronic form of BCWD following recovery from typical BCWD. It is characterised by erratic "corkscrew" swimming, blackened tails and spinal deformities.

The sweetfish which was infected with cold water disease.

In rainbow trout fry syndrome, acute disease with high mortality rates occurs. Infected fish may show signs of lethargy, inappetance and exopthalmos before death.

A presumptive diagnosis can be made based on the history, clinical signs, pattern of mortality and water temperature, especially if there is a history of the disease in the area. The organism can be cultured for definitive diagnosis. Alternatively, histology should show periostitis, osteitis, meningitis and ganglioneuritis.

Treatment

Quaternary ammonium compounds can be added to the water of infected adult fish and fry. Alternatively, the antibiotic oxytetracycline can be given to adults, fry and broodstock. To prevent the disease, it is necessary to ensure water is pathogen-free and that water hardening is completed effectively for eggs.

Beet Vascular Necrosis

Beet vascular necrosis and rot is a soft rot disease caused by the bacterium *Pectobacterium carotovorum* subsp. *betavasculorum*, which has also been known as *Pectobacterium betavasculorum* and *Erwinia carotovora* subsp. *betavasculorum*. It was classified in the genus *Erwinia* until genetic evidence suggested that it belongs to its own group; however, the name Erwinia is still in use. As such, the disease is sometimes called Erwinia rot today. It is a very destructive disease that has been reported across the United States as well as in Egypt. Symptoms include wilting and black streaks on the leaves and petioles. It is usually not fatal to the plant, but in severe cases the beets will become hollowed and unmarketable. The bacteria is a generalist species which rots beets and other plants by secreting digestive enzymes that break down the cell wall and parenchyma tissues. The bacteria thrive in warm and wet conditions, but cannot survive long in fallow soil. However, it is able to persist for long periods of time in the rhizosphere of weeds and non-host crops. While it is difficult to eradicate, there are cultural practices that can be used to control the spread of the disease, such as avoiding injury to the plants and reducing or eliminating application of nitrogen fertilizer.

A table beet infected with *Pectobacterium carotovorum* subsp. *betavasculorum*. Note the rings of black vascular tissue colonized by the rotting bacteria.

Hosts

Fodder beets, sugar beets and fodder-sugar crosses are all susceptible to infection by *Pectobacterium carotovorum* subsp. *betavasculorum*. Today most beet cultivars are resistant to the pathogen, however, isolates vary geographically, and some cultivars of beets are only resistant to specific isolates of bacteria. For example, the cultivar USH11 demonstrates resistance to both Montana and California isolates, whereas Beta 4430 is highly susceptible to the Montana isolates but resistant to the California isolate. Other cultivars resistant to California isolates of *Pectobacterium caratovorum* subsp. *betavasculorum* include Beta 4776R, Beta 4430R and Beta 4035R, but HH50 has been found to be susceptible.

Breeding for resistance to other diseases such as beet yellows virus without also selecting for vascular necrosis resistance can leave cultivars susceptible to the pathogen. For example, the use of USH9A and H9B in California's San Joaquin valley is thought to have led to an epiphytotic (severe) outbreak of disease in the early 1970s. This was likely because of the limited gene pool used when selecting strongly for resistance to beet yellows virus. Further information on resistant cultivars can be found in the section Management.

In addition to beets, *Pectobacterium carotovara* subsp. *betavasculorum* can also infect tomato, potato, carrots, sweet potato, radish, sunflower, artichokes, squash, cucumber and chrysanthemum. Other subspecies of *Pectobacterium carotovora* can also be pathogenic to beets. *Erwinia carotovara* subsp. *atroseptica* is a bacterial soft rot pathogen that is responsible for the disease Blackleg of Potato (*Solanum tuberosum*), and variants of this bacterium can cause root rot in sugarbeets,. This subspecies also has a wide host-range. *Erwinia carotovora* var. *atroseptica* has been detected in the rhizosphere of native vegetation and on weed species such as *Lupinus blumerii* and *Amaranthus palmeri* (pigweed). It is thought that the source of inoculums survives on these non-host plants in areas in which it is endemic as well as in the rhizosphere of other crops such as wheat and corn

Symptoms

Symptoms can be found on both beet roots and foliage, although foliar symptoms are not always present. If present, foliar symptoms include dark streaking along petioles and viscous froth deposits on the crown which are a by-product of bacterial metabolism. Petioles can also become necrotic and demonstrate vascular necrosis. When roots become severely affected, wilting also occurs. Below ground symptoms include both soft and dry root rot. Affected vascular bundles in roots become necrotic and brown, and tissue adjacent to necrosis becomes pink upon air contact. The plants that do not die completely may have rotted-out, cavernous roots.

Table beet stem infected with *Pectobacterium carotovorum* subsp. *betavasculorum*. Note entry through a wound.

Various pathogens can cause root rot in beets; however the black streaking on petioles and necrotic vascular bundles in roots and adjacent pink tissue help to distinguish this disease from others such as Fusarium Yellows. Additionally, sampling from the rhizosphere of infected plants and serological tests can confirm the presence of *Erwinia caratovora* subs.

Disease Cycle

Pectobacterium carotovorum subsp. *betavasculorum* is a gram negative, rod bacteria with peritichous flagella. For it to enter sugar beet, and thus cause infection, it is essential that there is an injury to the leaves, petioles or crown. Infection will often start at the crown and then move down into the root, and can occur at any point in the growing season if environmental conditions are favorable. Once the bacteria enters the plant, it will invade the vascular tissue and cause symptoms by producing plant cell wall degrading enzymes, like pectinases, polygalactronases, and celluases. This results in discolored or necrotic vascular tissue in the root, and the tissue bordering the vascular bundles will turn reddish upon contact with air. Following the infection of the vascular tissue, the bacteria reproduce as long as food resources are available, and the root begins to rot. There is significant variability in the type of rot – it can range from a dry rot to soft and wet rot – because of the multitude of additional microorganisms that may colonize the damaged tissue.

Upon death of the sugar beet, or harvest of the field, the pathogen appears to survive in select living plant tissue like beet roots, or volunteer beets. However, it does not appear to survive in sugar beet seeds, or live in the soil after harvest. It is also possible for the pathogen to infect injured carrots, potato, sweet potato, tomato, radish, sunflower, artichokes, squash, cucumber and chrysanthemums; however, since those are often planted in the same season as sugar beets, they are not likely to be overwintering hosts.

Environment

Injury to the leaves, petioles or crown is mandatory for the pathogen to gain entry to the host tissue. Accordingly, hail damage is correlated with a higher degree of disease outbreak. Young plants (less than eight weeks old) are also considered to be more prone to infection.

Temperature and availability of moisture are key factors in determining the rate of disease development. Warm temperatures, 25-30 °C, promote rapid disease development., and can result in acute symptoms. Symptoms are also reported to appear at temperatures as low as 18 °C, but disease development is slowed; below that temperature, infections do not develop. Excessive water also promotes disease development by providing a more optimal environment for the pathogen, and has been shown to be a key factor in augmenting disease outbreak in fields with sprinkler irrigation.

Agricultural

The degree of nitrogen fertilization is highly correlated to robust disease development: it has been shown that sugar beets supplied with excessive or adequate nitrogen are more diseased than sugarbeets with sub-optimal nitrogen levels. This is a paradox for farmers because, while increased nitrogen fertilization does increase sugar yield in non-infected sugarbeets, it also increases the severity of the disease if infection takes place. Thus, depending on the severity of infection, yield may go down with increased fertilizer use.

The spacing between plants also impacts the degree of infection: greater in-row spacing results in more diseased roots. This may be due to the fact that greater spacing promotes faster growth, and hence greater probability of cracks in the crown, or because of the increased amount of nitrogen available per plant.

Since the pathogen has multiple hosts, it is important for farmers to be wary of other plants in the surrounding area. It is possible for the pathogen to survive in weedy hosts, and can infect injured carrots, potato, sweet potato, tomato, radish, squash, and cucumber. Hence, the presence of these plants may increase the supply of inoculum.

Laboratory

If the pathogen is cultured in a lab, it can grow on Miller and Schroth media, can use sucrose to make reducing sugars, and can use either lactose, methyl alpha-glucoside, inulin or raffinose to make acids. It is also capable of surviving in culture medium sodium levels of up to 7–9%, and in temperatures as high as 39 °C.

Management

Since the bacteria cannot survive in seeds, the best way to prevent the disease is to ensure that vegetatively propagated plant material are clean of infection, such that the bacterium does not enter the soil. However, if the bacteria is already present, there are some methods that can be used to lessen the infection.

Cultural Practices

Because the bacteria readily enter the plant through wounds, management practices that decrease injury to the plants are important to control the spread of the disease. Cultivation is not recommended, as the machinery can become contaminated and physically spread the bacteria around the soil. Accidental leaf tearing or root scarring can also occur depending on the size of the crop, allowing the bacteria to enter more individual plants. If hilling the beets, great care must be taken to avoid getting soil into the crown, because the pathogen is soil-borne and this could expose the plant to more bacteria, thus increasing the risk of infection.

While most bacteria are motile and can swim, they cannot move very far due to their small size. However, they can be carried along by water, and a significant movement of *Pectobacterium* can be attributed to being carried downstream from irrigation and rainwater. To control the spread of the disease, limiting irrigation is another strategy. The bacteria also flourishes in wet conditions, so limiting excess water can control both the spread and severity of the disease.

Increased in-row spacing also causes more severe disease. In an infected field, yield decreased linearly when spacing was greater than 15 cm (6 in), so a spacing of 6 inches or less is recommended.

The bacteria can also utilize nitrogen fertilizer to accelerate their growth, thus limiting or eliminating the amount of nitrogen fertilizer applied will lessen the disease severity. For example, when fertilizer was applied to an infected field the infection rate per root increased from 11% (with no added nitrogen) to 36% (with 336 kg nitrogen/hectare), and sugar yields decreased.

Resistance

The bacteria can survive in the rhizosphere of other crops such as tomato, carrots, sweet potato, radish, and squash as well as weed plants like lupin and pigweed, so it is very hard to get rid of it completely. When it is known that the bacterium is present in the soil, planting resistant varieties can be the best defense against the disease. Many available beet cultivars are resistant to *Pectobacterium carotovorum* subsp. *betavasculorum*, and some examples are provided in the corresponding table. A comprehensive list is maintained by the USDA on the Germplasm Resources Information Network. Even though some genes associated with root defense response have been identified, the specific mechanism of resistance is unknown, and it is currently being researched.

Biological Control

Some bacteriophages, viruses that infect bacteria, have been used as effective controls of bacterial diseases in laboratory experiments. This relatively new technology is a promising control method that is currently being researched. Bacteriophages are extremely host-specific, which makes them environmentally sound as they will not destroy other, beneficial soil microorganisms. Some bacteriophages identified as effective controls of *Pectobacterium carotovorum* subsp. *betavasculorum* are the strains ΦEcc2 ΦEcc3 ΦEcc9 ΦEcc14. When mixed with a fertilizer and applied to inoculated calla lily bulbs in a greenhouse, they reduced diseased tissue by 40 to 70%. ΦEcc3 appeared to be the most effective, reducing the percent of diseased plants from 30 to 5% in one trial, to 50 to 15% in a second trial. They have also been used successfully to reduce rotting in lettuce caused by *Pectobacterium carotovorum* subsp. *carotovorum*, a different bacterial species closely related to the one that causes beet vascular necrosis.

While it is more difficult to apply bacteriophages in a field setting, it is not impossible, and laboratory and greenhouse trials are showing bacteriophages to potentially be a very effective control mechanism. However, there are a few obstacles to surmount before field trials can begin. A large problem is that they are damaged by UV light, so applying the phage mixture during the evening will help promote its viability. Also, providing the phages with susceptible non-pathogenic bacteria to replicate with can ensure there is adequate persistence until the bacteriophages can spread to the targeted bacteria. The bacteriophages are unable to kill all the bacteria, because they need a dense population of bacteria in order to effectively infect and spread, so while the phages were able to decrease the number of diseased plants by up to 35%, around 2,000 Colony Forming Units per milliliter (an estimate of living bacteria cells) were able to survive the treatment. Lastly, the use of these bacteriophages places strong selection on the host bacteria, which causes a high probability of developing resistance to the attacking bacteriophage. Thus it is recommended that multiple strains of the bacteriophage be used in each application so the bacteria do not have a chance to develop resistance to any one strain.

Importance

The disease was first identified in the western states of, California, Washington, Texas, Arizona and Idaho in the 1970s and initially led to substantial yield losses in those areas. *Erwinia caratovara* subsp *betavascularum* was not discovered in Montana until 1998. When it first appeared, beet vascular necrosis caused individual farm yield loss ranging from 5–70% in Montana's Bighorn Valley. Today, yield losses from the disease are generally infrequent and patchy as most producers plant resistant varieties. Infection rate is generally low if resistant cultivars are chosen; however, warmer and wetter conditions can lead to higher than normal instance of disease.

If infection does occur, bacterial root rots can not only cause economic losses in the field, but also can in storage and processing as well. In processing plants, rotten roots complicate slicing and the bacterially-produced slime can clog filters. This is especially problematic with late-infected beets which are generally harvested and processed along with healthy beets. The disease can also lower sugar-content which greatly reduces the quality.

References

- Ghosh, editors-in-chief, Thomas M. Habermann, Amit K. (2008). Mayo Clinic internal medicine: concise textbook. Rochester, MN: Mayo Clinic Scientific Press. p. 789. ISBN 978-1-4200-6749-1.
- Kabra, [edited by] Vimlesh Seth, S.K. (2006). Essentials of tuberculosis in children (3rd ed.). New Delhi: Jaypee Bros. Medical Publishers. p. 249. ISBN 978-81-8061-709-6.
- Ryan, Kenneth J.; Ray, C. George, eds. (2004). Sherris Medical Microbiology (4th ed.). McGraw Hill. pp. 451–3. ISBN 0-8385-8529-9. OCLC 61405904.
- Dolin, [edited by] Gerald L. Mandell, John E. Bennett, Raphael (2010). Mandell, Douglas, and Bennett's principles and practice of infectious diseases (7th ed.). Philadelphia, PA: Churchill Livingstone/Elsevier. pp. Chapter 250. ISBN 978-0-443-06839-3.
- McMenamin, Dorothy (2011). Leprosy and stigma in the South Pacific : a region-by-region history with first person accounts. Jefferson, N.C.: McFarland. p. 17. ISBN 978-0-7864-6323-7.
- Gibson, Peter G. (ed.); Abramson, Michael (ed.); Wood-Baker, Richard (ed.); Volmink, Jimmy (ed.); Hensley, Michael (ed.); Costabel, Ulrich (ed.) (2005). Evidence-Based Respiratory Medicine (1st ed.). BMJ Books. p. 321. ISBN 978-0-7279-1605-1.

Permissions

All chapters in this book are published with permission under the Creative Commons Attribution Share Alike License or equivalent. Every chapter published in this book has been scrutinized by our experts. Their significance has been extensively debated. The topics covered herein carry significant information for a comprehensive understanding. They may even be implemented as practical applications or may be referred to as a beginning point for further studies.

We would like to thank the editorial team for lending their expertise to make the book truly unique. They have played a crucial role in the development of this book. Without their invaluable contributions this book wouldn't have been possible. They have made vital efforts to compile up to date information on the varied aspects of this subject to make this book a valuable addition to the collection of many professionals and students.

This book was conceptualized with the vision of imparting up-to-date and integrated information in this field. To ensure the same, a matchless editorial board was set up. Every individual on the board went through rigorous rounds of assessment to prove their worth. After which they invested a large part of their time researching and compiling the most relevant data for our readers.

The editorial board has been involved in producing this book since its inception. They have spent rigorous hours researching and exploring the diverse topics which have resulted in the successful publishing of this book. They have passed on their knowledge of decades through this book. To expedite this challenging task, the publisher supported the team at every step. A small team of assistant editors was also appointed to further simplify the editing procedure and attain best results for the readers.

Apart from the editorial board, the designing team has also invested a significant amount of their time in understanding the subject and creating the most relevant covers. They scrutinized every image to scout for the most suitable representation of the subject and create an appropriate cover for the book.

The publishing team has been an ardent support to the editorial, designing and production team. Their endless efforts to recruit the best for this project, has resulted in the accomplishment of this book. They are a veteran in the field of academics and their pool of knowledge is as vast as their experience in printing. Their expertise and guidance has proved useful at every step. Their uncompromising quality standards have made this book an exceptional effort. Their encouragement from time to time has been an inspiration for everyone.

The publisher and the editorial board hope that this book will prove to be a valuable piece of knowledge for students, practitioners and scholars across the globe.

Index

N

O

P

R

S

T

V

www.ingramcontent.com/pod-product-compliance
Lightning Source LLC
LaVergne TN
LVHW080622210926
848485LV00003B/199
9781639874514